i

This Page Intentionally Left Blank

About Us

The Superintendent of the United States Military Academy (USMA) at West Point officially approved the creation of the Center for Nation Reconstruction and Capacity Development (C/NRCD) on 18 November 2010. Leadership from West Point and the Army realized that the United States Army, as an agent of the nation, would continue to grapple with the burden of building partner capacity and nation reconstruction for the foreseeable future. The Department of Defense (DoD), mainly in support of the civilian agencies charged with leading these complex endeavors, will play a vital role in nation reconstruction and capacity development in both pre and post conflict environments. West Point affords the C/NRCD an interdisciplinary and systems perspective making it uniquely postured to develop training, education, and research to support this mission.

The mission of the C/NRCD is to take an interdisciplinary and systems approach in facilitating and focusing research, professional practice, training, and information dissemination in the planning, execution, and assessment of efforts to construct infrastructure, networks, policies, and competencies in support of building partner capacity for communities and nations situated primarily but not solely in developing countries. The C/NRCD will have a strong focus on professional practice in support of developing current and future Army leaders through its creation of cultural immersion and research opportunities for both cadets and faculty.

The research program within the C/NRCD directly addresses specific USMA needs:
- Research enriches cadet education, reinforcing the West Point Leader Development Systems through meaningful high impact practices. Cadets learn best when they are challenged and when they are interested. The introduction of current issues facing the military into their curriculum achieves both.
- Research enhances professional development opportunities for our faculty. It is important to develop and grow as a professional officer in each assignment along with our permanent faculty.
- Research maintains strong ties between the USMA and Army/DoD agencies. The USMA is a tremendous source of highly qualified analysts for the Army and the DoD.
- Research provides for the integration of new technologies. As the pace of technological advances increases, the Academy's education program must not only keep pace but must also lead to ensure our graduates and junior officers are prepared for their continued service to the Army.
- Research enhances the capabilities of the Army and DoD. The client-based component of the C/NRCD research program focuses on challenging problems that these client organizations are struggling to solve with their own resources. In some cases, USMA personnel have key skills and talent that enable solutions to these problems.

For more information please contact:

> Center for Nation Reconstruction and Capacity Development
> Attn: Dr. John Farr, Director
> Department of Systems Engineering
> Mahan Hall, Bldg. 752
> West Point, NY 10996
> John.Farr@usma.edu
> 845-938-5206

This Page Intentionally Left Blank

Table of Contents

List of Figures

List of Tables

Chapter 1
Introduction

1.1 Problem Statement

The United States (US) Army and our other armed forces need a framework and metrics for quantifying the return on investment (ROI) of programs or projects tied to renewable energy and the security of its supply to military installations. Given the frailty of energy infrastructure at the national level, our ability to project military power in times of war, along with our entire economic and social well being, requires hardened, resilient, and redundant power systems[1] that can survive a host of cyber and physical attacks.

Currently our country is in a state of extreme dependence upon fossil fuels to support the stability of our economy. The government has not only begun to understand the damage that fossil fuels have on the environment, but also that this dependence places the US in a vulnerable position to a myriad of attacks and embargos. Thus, the Army enacted the Energy Security Implementation Strategy[2], and numerous renewable energy mandates, orders, and laws have been decreed that dramatically affect its energy and environmental policies. For example, the Army Energy Program website[3] contains 14 Army Guidelines, 4 Department of Defense (DoD) Guidances, 5 Presidential Orders, and 5 Federal Laws and Statutes that are referenced as affecting Army energy and environmental issues. Figure 1.1 illustrates how the modern installation manager and commander must navigate a complex atmosphere to ensure that their installation can conduct its mission both in war and peacetime.

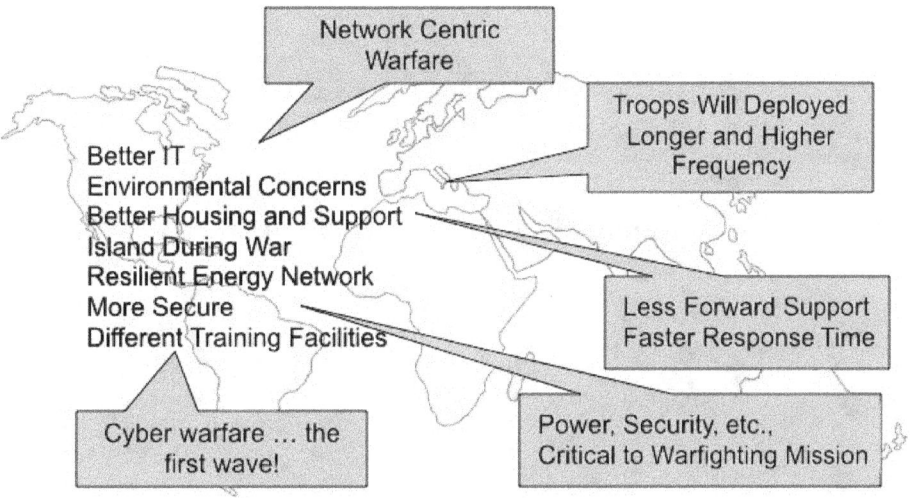

Figure 1.1 *Issues for the 21st century military installation*

If the US were attacked from either an external, sophisticated enemy or even an internal, disgruntled employee, the energy grid would be an easy target to damage/destroy that would lead to catastrophic results. This would cut off power to military bases and greatly degrade our ability to project force in addition to crippling the US economy. Figure 1.2 graphically shows how power is currently distributed to most military installations. Creative economics and engineering solutions will be needed to finance these upgrades needed not only to ensure the security of our installations, but to also comply with the myriad of

---

[1] When we use the word power systems we mean infrastructure, cyber, people, processes, etc.
[2] http://www.asaie.army.mil/Public/Partnerships/doc/AESIS_13JAN09_Approved%204-03-09.pdf accessed 14 November 2011
[3] See http://army-energy.hqda.pentagon.mil/policies/key_directives.asp accessed 14 November 2011

regulations, orders, and laws that govern how they are operated. Figure 1.3 shows the essential elements of the national energy grid.

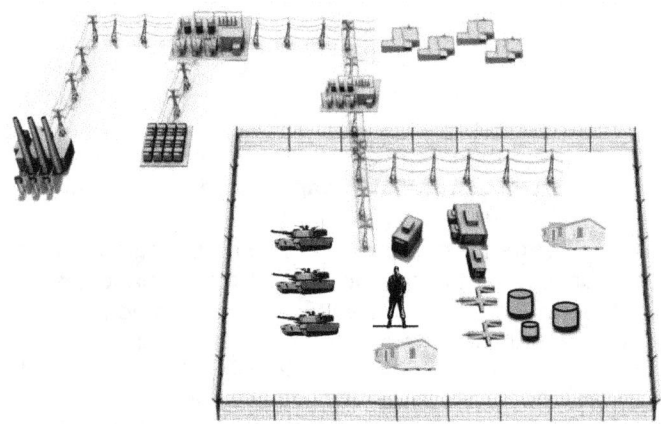

Figure 1.2 An installation that is dependent on external sources of power

Photo Removed Due to Copyright Restrictions

Renewable energy and the "going green" movement have been much discussed topics for years. For example, one major mandate for military installations was that by fiscal year 2010, renewable energy consumption must have been at a minimum of 5%. New buildings on military installations must have reduced their consumption of fossil-fuel-generated energy by 55% in 2010, and be completely free of it by 2030. Unfortunately, these goals currently lack the funding or engineering and technology to be achieved.

Some schools of thought call for military installations to have the ability to "island" themselves from the power grid in order to support their strategic mission. Ideally a military base would be able to switch off of the energy grid and still manage to provide power to the components of the installation essential to conduct their mission. This is important in the event of a physical or cyber attack on either the energy grid

---

[4] Taken from http://static.howstuffworks.com/gif/power-transmission.gif, accessed 14 November 2011

or on our energy resources.[5] Should this happen, the Army, or any service, must still be able to carry out its key functions until the power is reestablished. Currently the military is getting its energy from traditional commercial vendors. However, there is no current plan or funding to harden those facilities and the supplying grid. In order to island itself, an installation would almost certainly have to move some kind of energy source onto its own footprint. Whether this is a solar, wind, biomass, geothermal, or any other kind of renewable energy source, it should and must be economically viable when compared to fossil fuels. It has to be something that is not very vulnerable to attack (i.e., hardened), while at the same time, something that minimizes the amount of damage in the case that it is attacked (consider a nuclear facility). It has to be available all of the time (i.e., reliable) but take up as small of an amount of land as possible with minimal environmental impact. Critical national security and homeland defense missions are at risk of extended outage from failure of the grid. Currently the Army and DoD have implemented goals and policies to facilitate energy security and other NetZero[6] energy initiatives. However, from a solely ROI perspective these alternative energies are not cost efficient when compared to fossil fuels. The argument must be made that the value of alternative energy is directly related to security and environmental stewardship, offsetting the financial cost comparisons. In essence, we are trying to quantify the synergism between investments in energy security and NetZero projects/portfolios. This is the focus of this research.

In order for there to be any chance of renewable energy being a viable option in the near future for military installations, creative financing must be investigated and a means to quantify energy security are needed. When separate, these two programs (renewable energy and energy security) are financially infeasible; however, together they have a chance of being economically viable. If the energy system is successfully able to provide hardened, resilient, and reliable power systems, then its value will be greater than just the cost to procure and operate. The security component will support NetZero initatives. This would also allow installations to become NetZero and take care of the environmentally friendly mandates.

This research presents a framework for the Army and hopefully other services to use in order to evaluate the different renewable energy and energy security options by quantifying their value. Most military installations will require the use of a number of different energy sources working together simultaneously in order to make any of these options possible; thus, we will utilize a portfolio approach. Using this approach along with other analytical techniques, our goal is to:

1) Quantify the value of energy security and environmental investments,
2) Provide a transparent methodology for ranking projects, and
3) Quantify the overlap between renewable energy sources that serve to increase energy security and green energy initiatives in the Army

At the national level, energy security is a complex phenomenon incorporating a variety of economic, social, national security, and environmental aspects. For our work we are only focusing on the economics of energy security with regards to military installations. In reality, the social implications cannot be ignored. For example, what would be the response if a military installation had power and the local communities were without power for a significant amount of time?

Table 1.1 contains a threat matrix as a function of physical, cyber, and natural attacks. The table shows the consequence and the target. From that table you can see that 1) critical infrastructure is probably the most vulnerable to a wide range of threats and 2) terrorist and enemies are by far the most lethal.

---

[5] The social aspects of a military installation having power and much of the populace without power will not be addressed in this report. However, this is a real concern in the event of a natural or manmade disaster that would cripple the national energy grid.

[6] A Net Zero Energy Installation (NZEI) is an installation that produces as much energy on site as it uses, over the course of a year. To achieve this goal, installations must first implement aggressive conservation and efficiency efforts while benchmarking energy consumption to identify further opportunities. The next step is to utilize waste energy or to "re-purpose" energy. Boiler stack exhaust, building exhausts or other thermal energy streams can all be utilized for a secondary purpose. Co-generation recovers heat from the electricity generation process. The balance of energy needs then are reduced and can be met by renewable energy projects. Taken from http://army-energy.hqda.pentagon.mil/netzero/, accessed 14 November 2011.

Table 1.1 Threat matrix as a function of modes (modified from Ramirez-Marquez, 2007)

| Mode | Threat Matrix - Human | | | | | | | |
| | Consequence | | | Target | | | | |
	Threaten	Disrupt	Destroy/ Damage	Economy	People	Symbolic	Critical Infra-structure	Environ-ment
Physical - Intended								
Spies		✓	✓			✓	✓	
Terrorists	✓	✓	✓		✓	✓	✓	
Criminals	✓	✓		✓				
Vandals	✓	✓	✓			✓	✓	✓
Enemies		✓	✓	✓	✓	✓	✓	
Disgruntled	✓	✓	✓			✓	✓	✓
Physical - Unintended								
Software		✓					✓	
Human		✓					✓	
Organ		✓					✓	
Info Tech		✓					✓	
Hardware		✓					✓	
Cyber - Intended								
Hackers		✓				✓	✓	
Spies		✓	✓	✓		✓	✓	
Terrorists	✓	✓	✓	✓		✓	✓	✓
Criminals		✓		✓				
Vandals	✓	✓	✓			✓	✓	
Disgruntled	✓	✓		✓		✓		
Enemies	✓	✓	✓	✓		✓	✓	✓
Cyber - Unintended								
Software		✓		✓		✓	✓	
Human		✓		✓		✓	✓	
Organ		✓		✓		✓	✓	
Info Tech		✓		✓		✓	✓	
Hardware		✓		✓		✓	✓	
Natural								
Hurricanes	✓	✓	✓	✓			✓	✓
Earthquake		✓	✓	✓			✓	✓
Floods	✓	✓	✓	✓			✓	✓
Tornados		✓	✓	✓			✓	✓
Drought	✓	✓	✓				✓	✓
Fire	✓	✓	✓				✓	✓
Volcanoes		✓	✓				✓	✓

1.2 Interdependency Between Energy Security and NetZero

One of the derived benefits of energy security investments is that they also support NetZero initiatives. As shown in Figure 1.4, both energy security and NetZero investments are affected by many factors. Figure 1.5 is a Systemigram of the interrelationships between NetZero and energy security. Systemigrams have a multitude of uses in business, transportation, military operations, and for the most part any topic matter that requires graphical representations of system's sub systems and their interdependencies.

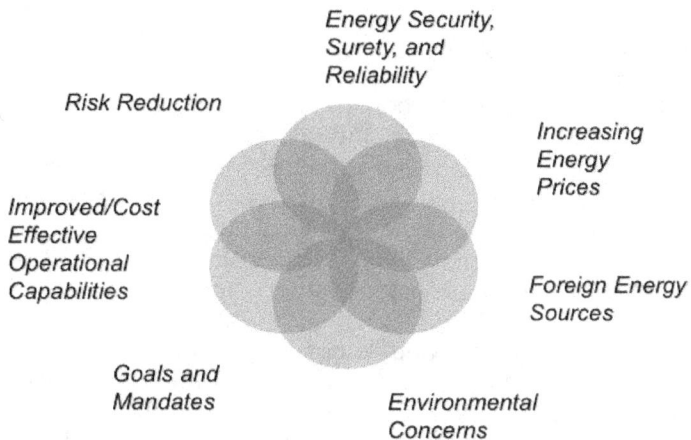

Figure 1.4 Factors affecting NetZero and energy security initiatives

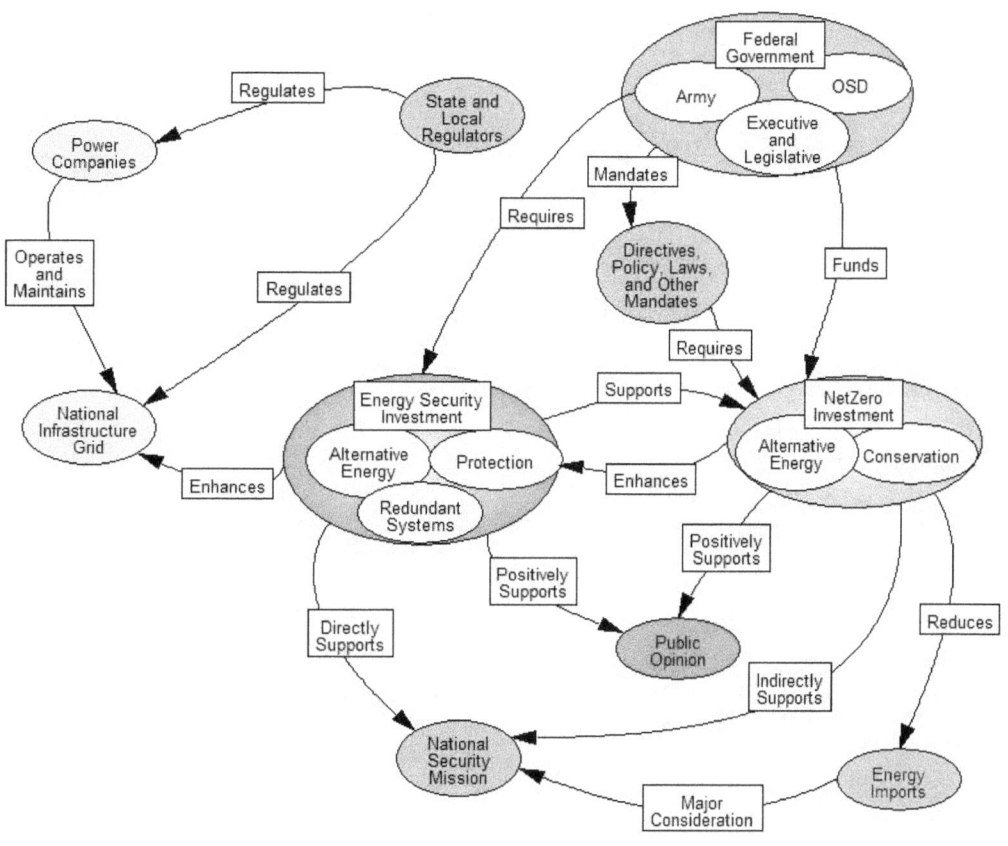

Figure 1.5 Systemigram showing the dependencies of NetZero and energy security

National military strategy for energy security is evolving. From a strategic perspective that are two distinct options for ensuring our military facilities can continue to operate in a major attack. First, the technology exists for military installations to island and essentially be self-contained. For example numerous companies have commercially available self-contained small nuclear reactors[7]. However, the social implications are not well understood of having military facilities with power when much of the populace would be without power for extended periods of time. The other option is to invest in the national power grid. The DoD could work with regional power authorities, regulators, developers of standards, invest research and development funding, etc., to help modify the grid to where is more hardened, reliable, and resilient. In reality the DoD has not chosen either of these paths. Instead the focus has been on reduced energy consumption, growing access to renewable power, smarter power management, etc. Also, from a national perspective Netzero has received more funding, political support, and visibility than the issue of energy security. If we can leverage the political will of investing in NetZero endeavours that will also support energy security solutions this is a win-win situation for the Army and the rest of the DoD.

Physical and cyber attacks to the energy grid and supporting elements (i.e., the world wide web, computer control systems, etc.) is a long-known risk. Open sources report has disclosed to public they have information of cyber attacks against power system controls from outside the US. Multi-city outages and rolling blackouts have become more common. Right now extortion and disgruntled employers are the attacked and not sophisticated enemies such as China. Our current nation power infrastructure is fragile and not resilient – in a time of war it would be an easy and crippling first target. We, the DoD, must be prepared to execute our national security mission. Without power this is not possible.

---

[7] See http://www.foxnews.com/story/0,2933,464625,00.html for an article that discusses at least three companies that are developing these roughly 50 MW reactors.

Chapter 2
Literature Review

2.1 Introduction

As our guiding principle, we have identified the five Strategic Energy Security Goals (ESGs) that the Army has adopted and include[8]:

- ESG 1. Reduced energy consumption,
- ESG 2. Increased energy efficiency across platforms and facilities,
- ESG 3. Increased use of renewable/alternative energy,
- ESG 4. Assured access to sufficient energy supplies, and
- ESG 5. Reduced adverse impacts on the environment.

Another way of spelling out the objectives of these goals is taken from a document on the implementation of Energy Security for the Army:

"Surety, Survivability, Supply, Sufficiency, Sustainability – these are the core characteristics defining the energy security necessary for the full range of Army missions. Energy security for the Army means preventing loss of access to power and fuel sources (surety), ensuring resilience in energy systems (survivability), accessing alternative and renewable energy sources available on installations (supply), providing adequate power for critical missions (sufficiency), and promoting support for the Army's mission, its community, and the environment (sustainability)." (Army Senior Energy Council, 2009)

In order to match our perspective in this report, you should consider the words "surety" and "survivability". If "surety" is the prevention of loss to access to power and fuel sources, then we must concern ourselves with energy all along its supply chain. For example, a fuel oil generator is not a viable generation method when we can't get petroleum to it because the supply convoys are constantly being hit. There is some uncertainty of the next shipment making it to its destination (lack of surety). A Forward Operating Base (FOB) with a self-contained nuclear reactor, on the other hand, would not be as susceptible to this uncertainty, as they would not require near as frequent shipments to be able to continue to provide power.

Secondly, the word "survivability" addresses not only the system's resilience against attack, but also the reliability of the system itself. This includes redundant systems that would allow the installation to continue to operate in the event of disruption of the primary means of energy generation, and the layers of security that one would have to breach in order to damage our infrastructure.

Developing quantifiable measures for surety, survivability, etc., is difficult. For example, in earlier work conducted along the same lines as our project's goal, security was measured by the number of "Family Days Saved" and "Staff Mission Days Saved". These measures were further explained as the "expected improvement" in respective days lost to power failures.[9] While this is technically a way of measuring energy security, in the sense that it is available for use 100 percent of the time, we must consider other threats to our supply. When we measure the security of energy on an installation, both physical and cyber protection should be considered. In this report, we discuss several of the alternative measures that we've considered for use in our modeling approach to solving the problem.

---

[8] Taken from the Army Energy Security Implementation Strategy, at
http://www.asaie.army.mil/Public/Partnerships/doc/AESIS_13JAN09_Approved%204-03-09.pdf, accessed 14 November 2011
[9] "A Value-Focused Approach to Justify the Cost of Energy Security" Slide 8, MORS Symposium

Methodology to Support Prioritization and Justification for
Energy Security and Renewable Energy Products

A resource that can be used to evaluate the security measures taken at any specific installation is the Security System Effectiveness Scale.[10] This scale aggregates the impact of a general kind of terrorist attack, given a certain level of security implementations. The more severe the potential for damage, the higher the threat level that is assigned to the installation.

In a study that examines security measures taken by the United States Navy, several methods of increasing a site's resilience against attacks are mentioned. Amongst these, several have the ability to be measured: standoff area surrounding buildings, the sophistication of barriers, and the sophistication of intrusion detection sensors. Standoff can be measured in linear distance, and the level of sophistication for different systems can be measured according to what generation they belong to. One point in particular is highlighted throughout the report: access control.[11] At the very least, a value could be assigned to the different generations of security devices. While the draft mentions several of these methods for increasing the security presence on an installation, the meanings of the changes are hard to quantify. For future study, another way of approaching this issue might be to rate the risk of any of these systems failing and then look at those alternatives with the lowest levels of associated risk.

While evaluating each of these individual metrics for security might be one way to approach solving this problem, we need to make sure that our methodology is portable, and able to be implemented at any Army post. This lends to the idea that our metrics for security must therefore be more general.

Given our interpretation of the term "energy security", we could even go so far to say that this includes economical effects. Energy prices are rising – especially where petroleum is considered – and whether or not the United States will be able to support the level of imports it is at now might become a consideration. However, the extrinsic analysis of cost, and whether it is a terminal factor in deciding what course of action to take, will be conducted elsewhere. Rather, as far as supply is considered, I would instead address the possibility of the supply being attacked or otherwise disrupted. Miles and miles of natural gas pipes are hard to defend, and oil is highly flammable. The second and third order effects of the sabotage of our energy system are concerning security issues.

Using these measures for security, what our project team aims to accomplish is to devise a system with which we can evaluate different portfolios for their value to the DoD. While few of the value measures for security carry a definite price tag, we hope to show that the more abstract benefits from added security are worth the investment. Combined Data Envelopment Analysis (DEA) and Multiple Objective Decision Analysis (MODA) will enable us to present options to our stakeholders that reflect the priorities that have been highlighted to us. Whether or not any of these options are viable solutions to the problem is for the Pentagon to decide; we simply attempted to have examined a full spectrum of alternatives.

2.2 Overview of Energy Assessment Techniques
Currently renewable energy is neither cost effective nor reliable enough to be a viable replacement for fossil fuels. A portfolio approach must be used because of technical, cost, environmental, social risks, and the difficulty of achieving the necessary level of production with a single source. In an article written at Massachusetts Institute of Technology's Elsevier journal for nuclear engineering, they explore this exact same concept of portfolio energy production, but at a larger scale. They consider the idea of combining the three major forms of energy production: fossil fuels, renewable sources, and nuclear energy. According to MIT, "new constraints and new technologies suggest that in many cases these energy sources must be tightly coupled to meet society's requirements" (Forsberg 2008). The major challenge with electrical energy is the variance it has over the course of a year. Even now, fossil fuel energy

---

[10] This idea that this resource might be useful in quantifying security measures in our study was engendered by a report drafted by Cadet Ali Chouhdry, USCC Company B, 1st Regiment, USMA. The report is titled "IMPROVING SECURITY AT THE UNITED STATES MILITARY ACADEMY". The information is further cited to be from *Critical Infrastructure and Key Resources Sector-Specific Plan* (Dept. of Interior, Defense pg 28).
[11] "Installations 2025 Study Report," Army Science Board

production struggles to keep up with peak energy requirement periods. If a portfolio concept were used, renewable sources would easily fill in where the primary energy sources are lacking. For example, in some areas, energy production peaks when the sun is hottest in order to run air conditioning. At this same time, solar power would be able to produce at its highest levels as well, effectively leveling the requirement on fossil fuels (Forsberg 2008). As of right now, most non-fossil fuel methods that could be used to create peak electrical power have limitations. It is hard to depend on a form of energy that relies on natural elements like wind and solar power. In the article, new forms of non-fossil fuel producers such as hydrogen are explored, but for the purposes of considering portfolios, the main direction of the article is finding a combination of renewable sources that effectively negate existing limitations. In other words, since most renewable sources are less cost effective then fossil fuels, a portfolio can allow the option of using renewable when they are most effective, and then switching to other sources when they are not.

2.3 Energy Security Evaluation Techniques

The literature contains many references to energy security especially from a national strategic perspective. For example the Nautilus Institute for Security and Sustainable Development (1998) presents a comprehensive review of the state of the art and makes the conclusion that there are no real analytical techniques for quantifying the value of energy security. Also, Hughes (2009) presents an Analytical Hierarchy Process very similar to the technique we are proposing. However, that work was mainly focused on national energy security.

The US Army Corps of Engineers came up with an analysis of the energy security on army installations. They discussed the information mentioned earlier in our report including the possibilities of "islanding". This report then lists the different alternatives that they will use before applying their criteria to them. However, when they get to this point, each alternative has three options for each criterion. Either it is "good" with that criteria, "no go", or is somewhere in the middle. This does not take into account the range that these alternatives can fall on. However, that they used a large number of alternatives is the best idea to come out of this report.

The Department of Energy (2006) has developed an energy security assessment guide. This guide is designed for federal installations to support:

- Initiation of the energy security assessment process,
- Vulnerability assessment,
- Energy preparedness and operations planning,
- Remedial action plans, and
- Management and implementation.

This report gives a detailed process for conducting vulnerability assessment planning, illustrated in Figure 2.1. However, it does not contain any type of methodology for assessing technologies that mitigate these vulnerabilities. The report does address another interesting concept; the synergism between security and environmental issues.

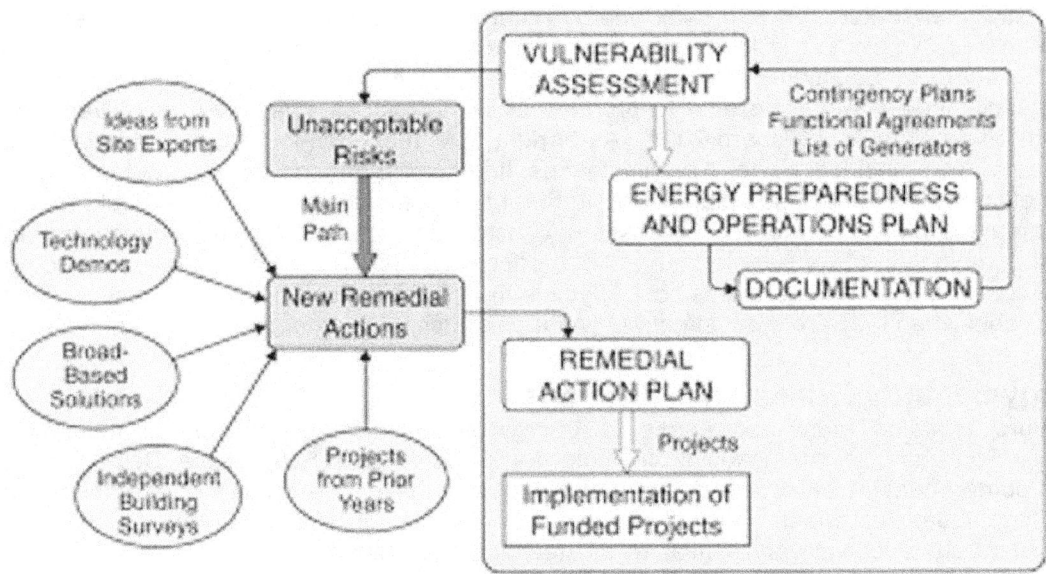

Figure 2.1 *Energy security program flow diagram modified to show the transition from the vulnerability assessment to remedial action*[12]

James Lambert, in his report titled "Energy Security of Military and Industrial Systems: Multicriteria Analysis of Vulnerability to Emergent Conditions including Cyber Threats", chooses to distinguish *emergent conditions* (conditions that may develop and affect investment decisions in the future) as either evidence-based or based upon the "subjective advocacy positions of the various stakeholders" (Part II: Background). The general approach of the report focused more on deciding which portfolio of investments best matches a given future scenario best summarized as: "a set of scenarios comprised of emergent and future conditions that influence energy security" (Part III: Technical Approach).

Lambert states that the results from this report present the stakeholders with a set of high performing alternatives, as well as a small set of scenarios that need to be more carefully studied. The difference that makes itself apparent between our report and theirs is that we are trying to maximize desirable achievement with one energy portfolio as dictated by the stakeholders. For instance, if we consider cyber threats to become more significant in the future, then we can increase our share in a portfolio that is more secure against them by increasing the weight of resistance to them. We have explored different future scenarios by generating alternative weights to use if the stakeholder is more concerned with achieving Netzero or energy security.

2.4 Value Hierarchy Models

At the request of the Deputy Assistant Chief of Staff for Installation Management (ACSIM) the Military Operation Research Society (MORS) conducted a workshop titled "A Value-focused Approach to Justify the Cost of Energy Security." The purpose of this workshop was to respond to questions asked by the ACSIM and included:

- How do we measure the value of investing in renewable and alternative energy at installations?
- What is the value that alternative energy provides in mission/energy assurance?

---

[12] Taken from Department of Energy, Federal Energy Management Program, Performing Energy Security Assessments — A How-To Guide for Federal Facility Managers, accessed at http://www1.eere.energy.gov/femp/pdfs/energy_security_guide.pdf, accessed December 15, 2011

The objective of the workshop was to demonstrate a framework to evaluate the value of improved mission/energy assurance in making energy investment decisions for installations. The recommendation from that workshop was to conduct a pilot project that applies and tests the framework.

This research, in essence, follows the recommendation of that workshop to conduct a pilot project. That workshop used the same value focused methodology we did. As shown in Figure 2.2, that work also aligned the value measures with the ESGs previously discussed.

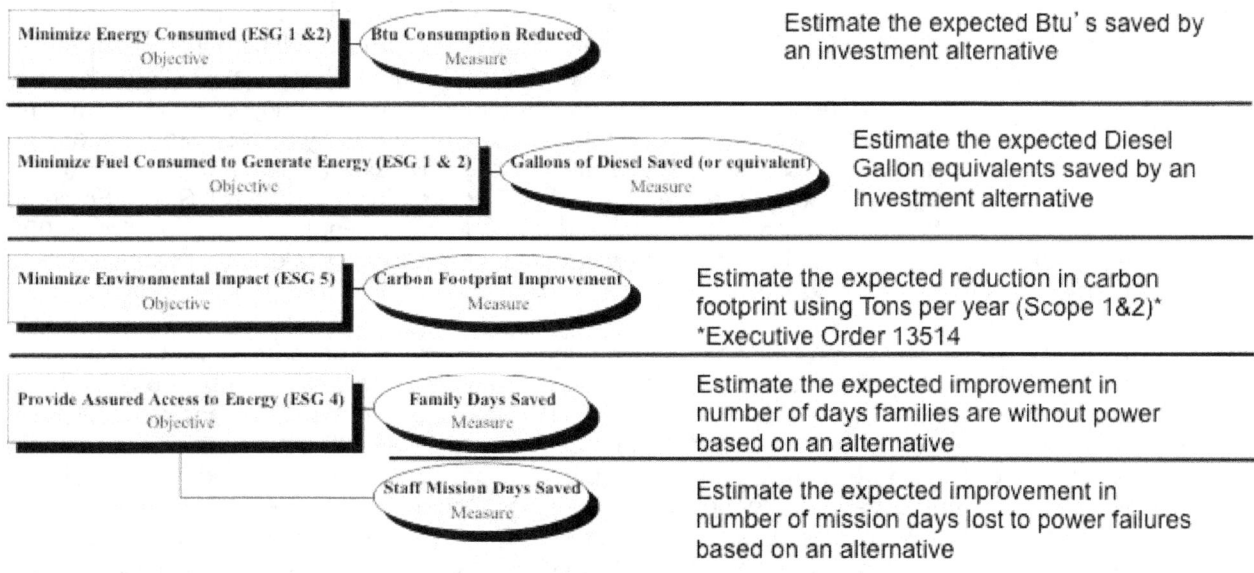

Figure 2.2 *Value focused thinking results from the MORs workshop*

Another important element of the workshop was conveying how values that are based upon an approved vision, policy, doctrine, etc., are key to developing a defensible model. From that workshop, it was noted that when developing value models, defensible and quantifiable measures are needed. We categorize data into four categories that include:
- Gold Standard
  - Based on an approved vision, policy, strategy, planning, or doctrine document;
  - Values have been thought about, discussed, and written down; and
  - Use work that has already been done and approved.
- Platinum Standard
  - Based on interviews with decision-makers and stakeholders
- Silver Standard
  - Uses data provided by stakeholder representatives
- Combined Standard
  - Combination of the above

2.5 Data Envelopment Analysis (DEA)
An increasingly popular way to evaluate performance is DEA as a quantitative tool. The basic concept for DEA is to take data from an existing entity or producer, referred to as the decision making unit (DMU), evaluate their performance, and then produce multiple possible alternatives (Cooper et al, 2000). There are a great amount of uses for DEA since it uses few assumptions and creates multiple outputs. As an "extreme point" method, DEA compares each measure of a producer with the best producer for that

measure. A simple DEA solution combines the best of all measures to produce "virtual best producers" (Anderson, 2011). DEA will be useful for analyzing energy security and renewable energy sources since there are already set producers, but not necessarily established portfolios that reach all the goals of NetZero or energy security goals. This review will cover the history, the uses and advantages of DEA, examples, and software used in DEA. There will also be a justification for the use of DEA in our current project with energy security and renewable energy products. Data Envelopment Analysis will help produce efficient portfolios and determine the best options for reaching NetZero and increase energy security.

The need for a new method of evaluation productivity was made apparent because previous attempts created very restrictive producers and didn't combine multiple inputs into an overall measure of efficiency. This marks the point when DEA entered into existence (Farrell, 1957). The initial DEA model was presented by Charnes, Copper, and Rhoades in 1978 in "Measuring the efficiency of decision making units" (Cooper et al, 2000). The first major use of the procedure was in Germany to estimate the marginal productivity of R&D (Brockhoff, 1970). Since DEA's creation there have been many journals and books written about its application. DEA was used more in the nineties; including the US Air Force in maintenance activities in different geographic locations and police forces in England (Cooper et al, 2000). Several types of DEA have been developed; for example, constant returns to scale and variable return to scale. The uses of DEA continue to expand.

Data Envelopment Analysis has found a place in product analysis because it uses fewer assumptions and combines inputs much better than other forms of analysis. DEA's ability to make comparisons is why it can now be found in a wide variety of fields. For example, in education, manufacturing, and retail chains it can be used to compare schools, branches, and departments, respectively. DEA is preferred in each field because it assists in identifying deficiencies and strengths in all products, not just providing one solution for all. DEA is a nonparametric method (Cooper et al, 2000). This means that there is no assumption that the structure of the model is fixed. The reason DEA is able to avoid this assumption is mainly due to the fact that the basic concept behind data envelopment is combining the strengths of multiple producers already created. This combination creates a new function referred to as the frontier. The frontier consists of all possible hybrids or "virtual" solutions. So, in this regard, DEA is able to take multiple inputs and create multiple outputs. Therefore, it is commonly used alongside another form of value modeling that helps highlight the best solutions from the multiple outputs. Another advantage is the ability to continue work with any form of unit. As long as the measures for each product are consistent, those measures can be any type of units. The frontier would then still maintain the same units in the virtual solutions.

A very basic example will help illustrate the advantages of DEA, and help see the efficiency frontier it creates that helps build multiple outputs. The example is a linear program, but it is important to realize that DEA is not limited to linear producers. The example is using baseball players created by Tim Anderson. Imagine you are a manager of a baseball team and need to find players with the ultimate goal of winning. Currently you have three possible players (producers). Each with specific statistics listed below in Table 2.1. You are currently looking at Home runs and Singles hit during 100 at-bats. Both are great for helping win.

Table 2.1 Example measures used for data envelopment analysis

Player	Singles (in 100 at-bats)	Home runs (in 100 at-bats)
A	40	0
B	20	5
C	10	20

There is currently no clear choice. Player C has a lot of home runs, player A has a much better batting average, and Player B is somewhere in the middle. How can this information now be used to help determine what player to pick? Each measure helps create an efficiency frontier. From the statistics, you are aware that separately it is possible to reach a maximum of twenty home runs or forty home runs. Linear programming then sets the relationship between the two measures, and an efficiency frontier is created, shown in Figure 2.3.

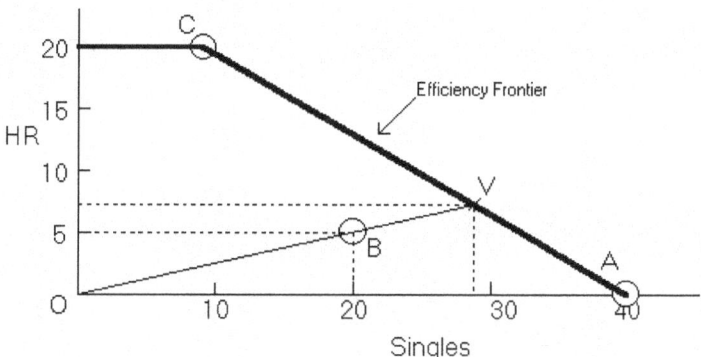

Figure 2.3 *Graphical representation of linear program used to determine efficiency in baseball example (Anderson, 2011)*

Using the relationship you would either choose one of the players already on the frontier or new incoming players that also reach the frontier. For example, a player with 25 singles and 10 homeruns is just as useful since it also lies on the efficiency frontier.

$$\lambda_Y = [0.5 * 40 + 0.5 * 10, 0.5 * 0 + 0.5 * 20] = [25, 10]$$

Just like in this example, any producer can be compared to the extreme points created by other producers.

Data Envelopment Analysis will be used for our project in order to identify portfolios that can reach all our goals. Each portfolio will have strengths that help create an efficiency frontier. This frontier will then be used to find the best hybrids. DEA has been used since the seventies in order to compare products. It has the advantage of using fewer assumptions then other forms of analysis and can take multiple inputs to create multiple outputs.

2.6 Summary

Currently, the government is unsure of how to attack the energy issues in this country. MODA and DEA give them the ability to compare the portfolios. MODA finds the value of each portfolio based on the stakeholder's needs and the actual figures that the portfolios produce. This allows the portfolios to be compared objectively as to which portfolios provide more value to the stakeholders than the others. With this, their values are compared to the cost which is graphed. From here the stakeholder is able to see how much it would cost for certain amounts of value. At the same time, DEA will be used and instead of trying to find value, DEA looks for efficiency. DEA will be used to find what combination of producers is most efficient in order to create the best portfolio possible. Using these two methodologies, their results will be compared and contrasted in order to find what most suits the stakeholder's needs.

This Page Intentionally Left Blank

Chapter 3
Model Development

3.1 Methodology

3.1.1 Multi-objective Decision Analysis Approach

Multi-objective Decision Analysis, or MODA, ranks alternatives to assist in selection of the preferred alternative. Specifically, it is useful in enhancing decision making for allocation of resources and solidifying support for a particular portfolio of projects. Using the objectives we obtained from various energy security and environmental requirements documents, this methodology is well suited to portfolio prioritization and optimization. The model will help to identify an appropriate mix of projects to maximize overall value.

It is important to first identify what is meant by the term "portfolio". A portfolio may be viewed at two levels. At the upper level, there exists an overall portfolio of projects that is comprised of the lower level of individual portfolios of projects. This lower level is the mix of projects from each of the stakeholders involved.

The MODA process begins with the development of value hierarchy similar to Figure 3.1. The three core functions and sub-functions are further broken down into objectives. The objectives identified can be further broken down into evaluation criteria in the value hierarchy model.

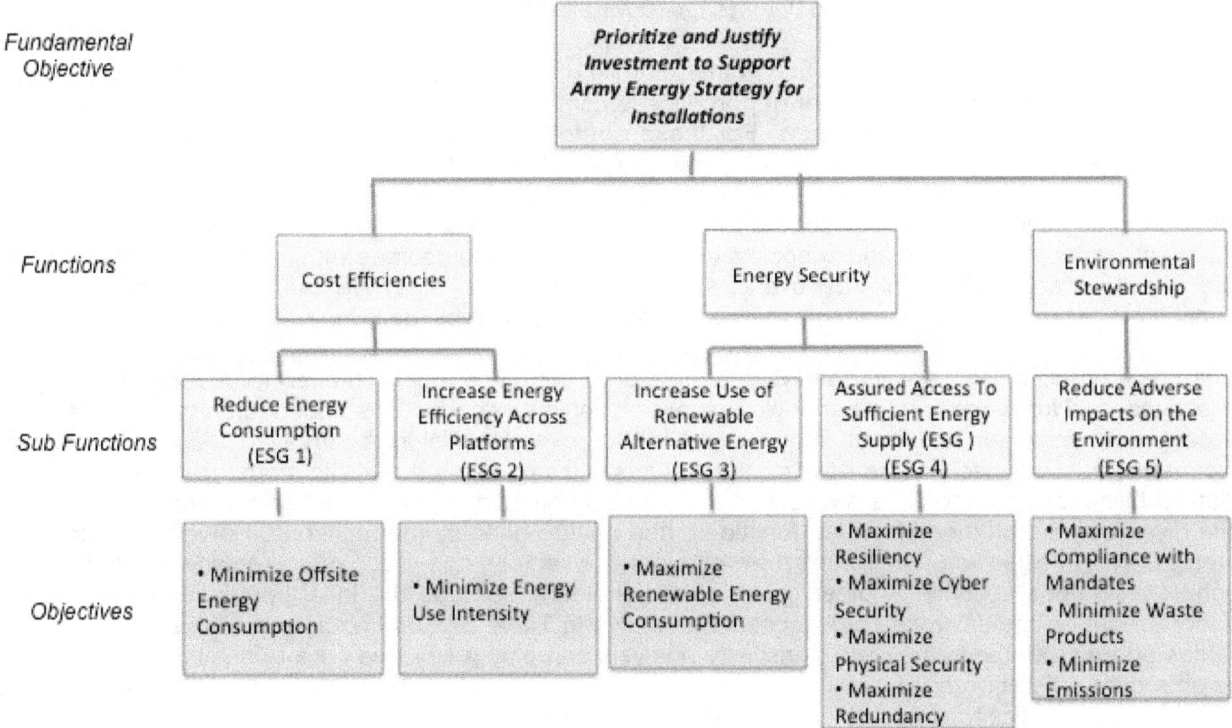

Figure 3.1 *Fundamental objective and functions for the energy strategy model*

As the evaluation criteria are dependent upon stakeholder analysis, they must be assigned local weights in the value hierarchy model. Appropriate evaluation criteria and local weights will be determined based on current information. We used mainly the information presented in Chapter 3 to develop our functions for assessing projects and portfolios.

Multi-objective decision or value analysis (Kirkwood, 1997) uses an overall value function which combines the multiple evaluation measures into a single measure of the overall value of each evaluation alternative, or portfolio of projects. Thus, different mixes of projects in a portfolio may be compared to determine the appropriate mix for maximizing value. Multi-objective value analysis is useful for structuring the judgments used in assessing the value of projects that comprise a portfolio in an organization with multiple and conflicting objectives. Multi-objective value analysis methods are based upon structured objectives, evaluation measures, value functions, and weights.

A multiple criteria value function based upon weights and scores are used to rank alternatives. An additive value function is used for this research since it is common (see Keeney and Raiffa, 1993). The additive multi criteria function V(a$_i$) can be expressed as

$$V(a_i) = \sum_{k=1}^{m} W_k V_k(a_i) \tag{4.1}$$

$$\text{where } \sum_{k=1}^{m} W_k = 1$$

and $0 \leq v_k(a_i) \leq 10$ for all k = 1, ... M.

The quantity $v_k(a_i)$ is the assessed value of the portfolio a_i. The weights W_k represent the tradeoffs across the criteria (weight and values). A set of portfolios is constructed and defined P={p$_1$, ...p$_n$} and used described the various energy solutions. For these portfolios we are interested how security, efficiency, regulations, etc., change how the portfolios or alternatives are scored.

When weights have been determined for the current situation, the model can be used to find the right mix of projects to maximize value and support a combination of core outcomes within a fixed budget portfolio. The mix of projects with the highest overall score adds the most value. We can then view projects as a function of cost or some other variable to make logical and defensible decisions.

When using multi-objective value analysis, a structured approach must be taken to develop the weights, objectives, and functions. In this paper we present objectives and functions based upon the experience of the authors, literature, and input from subject matter experts. We then surveyed a focus group with experience in energy security to develop the weights. This provides a realistic model to demonstrate the utility of this approach. Ideally, stakeholders should be involved at all levels. In general, there is often very little disagreement on the objectives, functions, and how to quantify the functions. However, stakeholder interests are reflected when assigning the weights. For example, one group of stakeholders might place a higher value upon security. Another group of stakeholders, such as the local populace, would place a higher weight on the environmental impacts, as shown in Table 3.1. Stakeholder buy-in is critical with all parties agreeing to the framework. Sensitivity analysis can play a key role here to show how varying the weights over different ranges can have little or major impact on the objective function. Table 3.2 shows the weights are typically assigned to the swing weight matrix.

Table 3.1 Swing weight distribution

		Importance of the value measure to the decision		
		High*	Medium	Low
Range of variation of the value measures	High	A	B₂	C₃
	Medium	B₁	C₂	D₂
	Low	C₁	D₁	E

Weights in the following cells need to follow these relationships (Parnell et al, 2008):
- A > all other cells
- B1 > C1, C2, D1, D2, E
- B2 > C2, C3, D1, D2, E
- C1 > D1, E
- C2 > D1, D2, E
- C3 > D2, E
- D1 > E
- D2 > E

Table 3.2 Swing weight values

		Importance of the Value Measure to the Decision Makers and Stakeholders		
		High	Medium	Low
Variation in Measure Ranges	High	100	70	40
	Medium	80	50	20
	Low	60	30	10

3.2 Summary

MODA is an effective tool for analyzing different projects and showing the stakeholder their value. The set process of creating alternatives, scoring them based off of stake holders needs, and assigning value gives an accurate portrayal of the return on any investment. This methodology can be used to analyze both NetZero and energy security goals.

Chapter 4
Demonstration Study

4.1 Introduction

In order to investigate the utility of our methodology, we developed a simple spreadsheet with which to implement our measures. The model contained herein is functional as a demonstration of the methodology, but certainly is in need of further development to include:

- Validation of weighting and scoring schema,
- Visual basic programming to present a more user-friendly model, and
- A more practical demonstration, with data tailored to the spreadsheet to validate its utility and to develop recommendations for improvement.

In order to demonstrate the methodology, we chose to use data collected at Fort Carson because the installation had recently funded an assessment of energy needs with the goal of achieving NetZero status. This study was conducted by the Department of Energy's National Renewable Energy Laboratory or NREL (see NREL, 2010). This report identified energy needs and provided economic data, potential technical solutions, and the information we needed to conduct a realistic demonstration study. However, the report had one significant limitation in that it provided renewable energy solutions without consideration of storage and other infrastructure needed to actually use the energy.

4.2 Energy Security Investment Strategy Model

A significant investment must be made in identifying desired objectives and quantifying their importance, also known as stakeholder analysis. The value hierarchy shown in Figure 4.1 begins with the fundamental objective at the top similar to an organizational chart. Identifying the fundamental objective is paramount to solving the right problem. The next level of the value hierarchy is the primary functions associated with the fundamental objective. We chose five functions of broad categories that must be addressed in order for us to solve the fundamental objective. They represent what must be done to accomplish the fundamental objective. Finally, we have the objectives that support each one of these functions. Each objective is how we intend to accomplish each respective function and is shown in Figure 4.1 how it is overlaid with the Army's energy security goals. The value model is where we capture the importance of each function and objective by using quantifiable value measures. The strength of MODA is the involvement of stakeholders in developing the criteria and the weights.

The value model is where we capture the importance of each function and objective and does not champion any one particular technique.

4.2.1 Reduce Energy Consumption Sub Function
Minimize Offsite Energy Consumption Objective

The amount of energy consumed will be measured by using the amount of site energy in lieu of source energy[13] used on the entire installation. For energy security we assume that offsite energy exposes that installation to the fragility of the national energy grid. In 2009, Fort Carson used 1.5 MMBtu (million

---

[13] **Site energy** is the amount of heat and electricity consumed by a building as reflected in utility bills. Site energy may be delivered to a facility in one of two forms: primary and/or secondary energy. **Primary energy** is the *raw fuel* that is burned to create heat and electricity, such as natural gas or fuel oil used in onsite generation. **Secondary energy** is the energy product (heat or electricity) created from a raw fuel, such as electricity purchased from the grid or heat received from a district steam system. A unit of primary and a unit of secondary energy consumed at the site are not directly comparable because one represents a raw fuel while the other represents a converted fuel. Therefore, in order to assess the relative efficiencies of buildings with varying proportions of primary and secondary energy consumption, it is necessary to convert these two types of energy into equivalent units of raw fuel consumed to generate that one unit of energy consumed on-site. To achieve this equivalency, the EPA uses the convention of source energy. from http://www.energystar.gov/index.cfm?c=evaluate_performance.bus_benchmark_comm_bldgs#diff, accessed May 16, 2012

British Thermal Units) in site energy. This value function is shown in Figure 4.2.

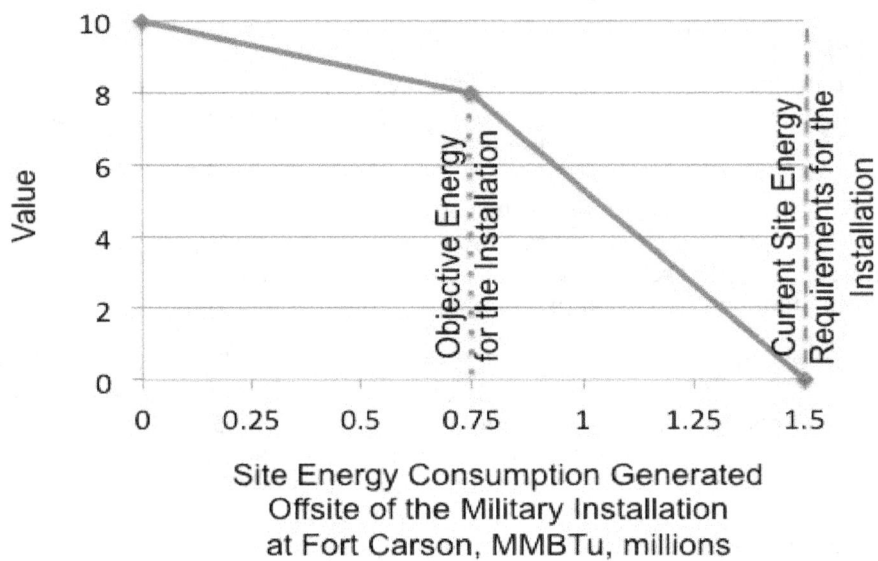

Figure 4.1 *Graphical representation for offsite energy consumption versus value*

4.2.2 Increase Energy Efficiency Across Platforms
Minimize Energy Use Intensity Objective

Energy use will be measured using energy use intensity (EUI), which is calculated by energy consumed per year divided, by total floor space (kBTu/ft^2). This can be expressed as percent of the average EUI for a similar building. Table 4.1 contains some rules of thumb for typical EUI values (NREL, 2010):

Table 4.1 *Typical EUI values*[14]

Building Type	Average EUI
Dormitory	151
Hospital	468
Hotel	228
K-12 School	169
Medical Office Building	134
Office Building	193
Retail Store	173

Note that EUI values are presented in kBtu/ft^2

Outside of a hospital, most buildings fall between 130-200 kBTu/ft^2. Similar to the way that other goals are being met, this will produce a value of 0. However, with the improvement of energy use, it is more than possible to reach their goals. Figure 4.3 presents the projected EUI through 2016 (NREL, 2010). The goal is to have the EUI under 90 by 2016 but is only currently projected to be around 120. In addition to the projected EUI, Figure 4.3 also shows the EUI and the corresponding value graph.

---

[14] Taken from http://www.energystar.gov/index.cfm?fuseaction=buildingcontest.eui, accessed 13 October 2011

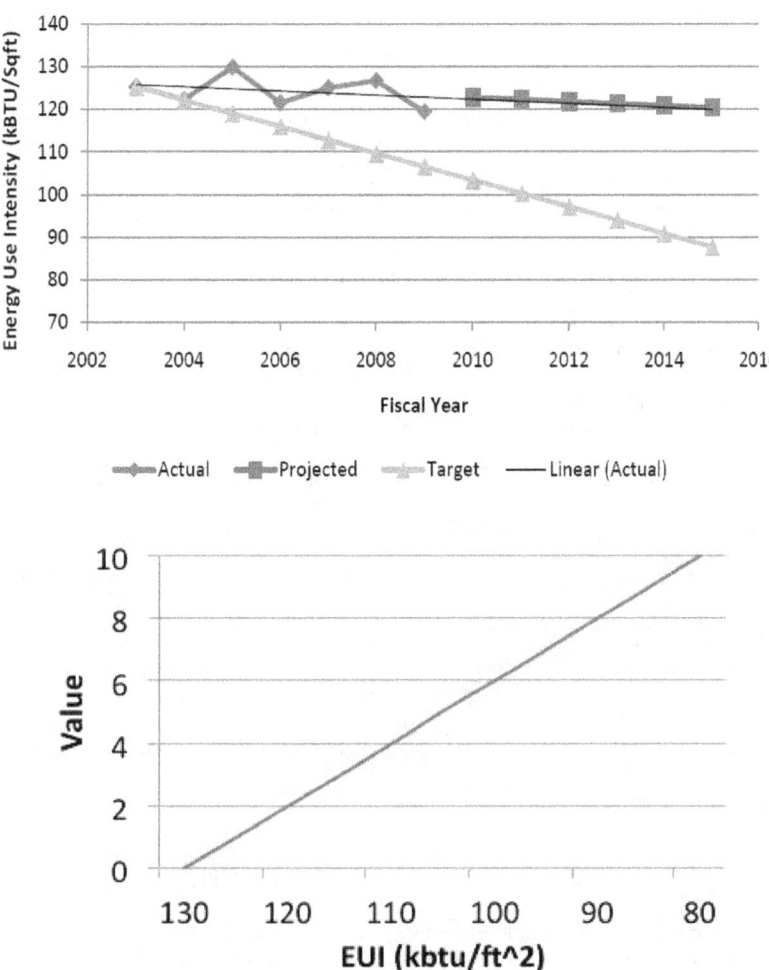

Figure 4.2 Graphical representation of the current state of EUI along with the values

Remember that we choose to look at portfolios in lieu of specific projects. Energy use intensity would certainly be relevant for retrofitting new controls, energy conversation measures, etc.

4.2.3 Reduce Adverse Impacts on the Environment
Maximize Compliance With Mandates Objective

A near endless list of laws and regulations exists in regards to the issues we are addressing in our study. These mandates set a variety of goals and standards for military installations to meet. Tables 4.2 and 4.3 below lists all of those mandates that apply. The ability of military installations to be able to conform to these mandates is important because it establishes their legitimacy as an efficient use of resources. The Army cannot afford to be viewed as a burden to the American taxpayer, and streamlining its installations in accordance with the standards outlined in the following table is one step to ensuring that it doesn't happen. Table 4.3 shows the three statutes that are currently mandated to the US Army. The table breaks down which statutes have target dates and what the goal of each statute is. As of right now the military is not meeting any of these goals and will not for the foreseeable future due to a lack of funding.

Table 4.2 *Recent regulations, laws, and orders that govern energy on an installation and their significance*

NetZero Significance Key	
1	Significant NetZero initiative
2	Minor NetZero initiative
3	Modern common building practices or Not applicable

NetZero Significance	Regulation, Law or Order
1	Federal Building Performance Standards: Minimum 30% more energy efficient than current ASHRAE or IECC standards; and Sustainable design principles are applied to the siting, design, and construction of all new and replacement buildings, EPAct05, sec 109 [Public Law 109-58]; 10 CFR 433; 10 CFR 435
2	High Performance and Sustainable Building Guidance: i) new construction and major renovations must comply with guidance and ii) 15% of existing building inventory by end of FY2015 incorporates outlined sustainable practices. EO 13423, sec 2(f)
3	Energy Efficient Products: Federal procurement of only Energy Star or FEMP designated products, EISA07, sec 525 [Public Law 110-140]
3	Meters: Electric, natural gas and steam meters installed, EPAct05, sec 103[Public Law 109-58] & EISA07, sec 434 [Public Law 110-140]
2	Solar Hot Water: 30% of hot water demand must be from solar hot water heaters, EISA07, sec 523 [Public Law 110-140]
1	Renewable Energy: consumption must be a minimum of 3% in FY07-09; 5% in FY 10-12; 7.5% in Fy2013+EPAct05, sec 203 [Public Law 109-58]
1	Renewable Energy Consumed: 50% of should come from new renewable sources and implement RE generation projects on the installation, EO13423, sec 2(b)
1	Fossil Fuels: New buildings must reduce fossil fuel-generated energy consumption, with 2003 as baseline, by 55% in 2010 and 100% by 2030, EISA07, sec 433 [Public Law 110-140]
3	Tax Incentives are transferred to the Designer for Federal Buildings; $0.30-$1.80 per square foot, depending on technology and amount of energy reduction. http://www.dsireusa.org/library/includes/incentive2.cfm?Incentive_Code=US40F&State=federal&currentpageid=1&ee=1&re=1, H.R. 1424: Div. B, Sec 303 (The Energy Improvement and Extension Act of 2008); 26 USC § 179D. Energy efficient commercial buildings deduction.
3	Energy Efficient Light bu bs: establishes energy efficiency standards for general service incandescent lamps. Bans most incandescent bu bs by 2012, EISA2007, sec 321 [Public law 110-140]
3	Life Cost Cycle Analysis: This Section amends 42 USC 8254(a)(1) to change 25 years to 40 years; EISA 07 Sec 441 [Public Law 110-140]; 10 CFR 436
3	Federal Building Performance Standards: Section 109 of EPAct 2005 required new federal buildings to be designed 30% below ASHRAE standards or IECC, to the extent that technologies employed are life-cycle cost-effective, and Sustainable Design Principles, EPAct05, sec 109 establishes"(i) if life-cycle cost-effective for new Federal buildings— "(II) sustainable design principles are applied to the siting, design, and construction of all new and replacement buildings;" Sustainable Design Principles (SDD) are those outlined by the DoA, USACE, DoE, EPA, and on the WBDG W3 site (http://www.wbdg.org/index.php) .
2	High Performance and Sustainable Building: EO 13423, sec 2(f), states that design principles must be applied to new and replacement buildings. All agencies must identify new building projects in their budget requests and identify those that meet or exceed the standard. To help achieve these energy reductions, new construction and major renovation of agency buildings must comply with the "Guiding Principles for Federal Leadership in High Performance and Sustainable Buildings" set forth in the *Federal Leadership in High Performance and Sustainable Buildings Memorandum of Understanding (2006)*, (High Performance and Sustainable Buildings Guidance, Final, 12/01/08 http://www.wbdg.org/pdfs/hpsb_guidance.pdf -) in addition to the energy goals and standards established by the federal Energy Policy Act of 2005. These building standards include a target energy use of 30% below the average building performance for new buildings, and a target that is 20% below the average for renovations.

Table 4.2 Recent regulations, laws, and orders that govern energy on an installation (cont)

3	Energy Efficient Products: Section 104 of EPAct 2005 directed federal agencies to purchase Energy Star and FEMP-designated products when procuring energy-consuming items covered by the Energy Star program, except when purchasing such items is not cost-effective or does not meet functional requirements of the agency. Agencies must also incorporate energy-efficient specifications in procurement bids and evaluations, and must only purchase premium efficient electric motors, air conditioning and refrigeration equipment. EPAct 2005 also instructed the General Services Administration (GSA) and the U.S. Department of Defense to clearly identify and display Energy Star and FEMP-designated products in any inventory, catalog or product listing.
3	Meters: EPAct05, sec 103, "(e) METERING OF ENERGY USE.—"(1) DEADLINE.—By October 1, 2012, in accordance with guidelines established by the Secretary under paragraph (2), all Federal buildings shall, for the purposes of efficient use of energy and reduction in the cost of electricity used in such buildings, be metered. Each agency shall use, to the maximum extent practicable, advanced meters or advanced metering devices that provide data at least daily and that measure at least hourly consumption of electricity in the Federal buildings of the agency.
2	Solar Hot Water: Section 523 of the EISA 2007 requires that at least 30% of the hot water demand for each new federal building or existing federal buildings undergoing a major renovation be met through the use of solar hot water heating, if it is determined to be life-cycle cost-effective.
1	Renewable Energy Consumption: The Energy Policy Act of 2005 established green power purchasing goals for the federal government, whereby the 7.5% of electricity used by federal agencies must be obtained from renewable sources by 2013, 3% in FY07-09, and 5% in FY10-12. Executive Order 13423 now requires at least half (50%) of the required renewable energy consumed by an agency in a fiscal year to come from sources placed in service in 1999 or later and to the extent possible, the agency implements renewable power generation projects on agency property for agency use.
2	Fossil Fuels: EISA07, sec 433. For new buildings or building undergoing major renovations requiring a GSA prospectus to Congress or at least $2.5 million, fossil fuels use to be reduced as compared to a similar building's use in FY 2003; percentages may be adjusted downward and sustainable design principles shall be applied. 55% by 2010; 65% by 2015; 80% by 2020; 90% by 2025; and 100% by 2030.
3	Tax Incentives: H.R. 1424: Div. B, Sec 303 (The Energy Improvement and Extension Act of 2008); 26 USC § 179D. Energy efficient commercial buildings deduction. A tax deduction of $1.80 per square foot is available to owners of new or existing buildings who install (1) interior lighting; (2) building envelope, or (3) heating, cooling, ventilation, or hot water systems that reduce the building's total energy and power cost by 50% or more in comparison to a building meeting minimum requirements set by ASHRAE Standard 90.1-2001. Energy savings must be calculated using qualified computer software approved by the IRS. In the case of energy efficient systems installed on or in government property, tax deductions will be given to the person primarily responsible for the systems' design. Deductions are taken in the year when construction is completed.
3	Energy Efficient Lighting: EISA2007, sec 321, Subtitle B—Lighting Energy Efficiency, Sec. 321. Efficient Light Bulbs. The Act establishes energy efficiency standards for general service incandescent lamps by modifying applicable sections of the Energy Policy and Conservation Act. Starting January 1, 2012, all general-service lamps must prove a minimum CRI, general service incandescent lamps must prove a minimum efficiency, and some incandescent lamps cannot exceed a maximum wattage.
3	Life Cost Cycle Analysis: [EISA 07 sec 441. PUBLIC BUILDING LIFE-CYCLE COSTS. Section 544(a)(1) of the National Energy Conservation Policy Act (42 U.S.C. 8254(a)(1)) is amended by striking "25" and inserting "40".): Establishment of life cycle cost methods and procedures – The Secretary, in consultation with the Director of the Office of Management and Budget, the Secretary of Defense, the Director of the National Institute of Standards and Technology, and the Administrator of the General Services Administration, shall– (1) establish practical and effective present value methods for estimating and comparing life cycle costs for Federal buildings, using the sum of all capital and operating expenses associated with the energy system of the building involved over the expected life of such system or during a period of 40 years, whichever is shorter, and using average fuel costs and a discount rate determined by the Secretary; and (2) develop and prescribe the procedures to be followed in applying and implementing the methods so established.
3	Institute of Standards and Technology, and the Administrator of the General Services Administration, shall– (1) establish practical and effective present value methods for estimating and comparing life cycle costs for Federal buildings, using the sum of all capital and operating expenses associated with the energy system of the building involved over the expected life of such system or during a period of 40 years, whichever is shorter, and using average fuel costs and a discount rate determined by the Secretary; and (2) develop and prescribe the procedures to be followed in applying and implementing the methods so established.

Table 4.3 The laws with basic information

	EPAct Section 203	Executive Order 13423	National Defense Authorization Act (2007)
Target/Goal	Increasing targets reaching 7.5% renewable content of electricity consumed	At least 7.5% of electric energy from new renewable energy with at least 50% from new renewable sources (after 1998)	25% of all energy consumed from renewable sources of supply
Target Dates	2013	2013	2025
Mandatory?	Yes	Yes	No
Considers thermal energy "renewable?"	No	Yes	Yes

These laws cover a variety of aspects all across installations, and as such, the method in which they are measured is different for most every mandate. Therefore, each mandate will be designated as one of three categories: significant NetZero initiative, minor NetZero initiative, or modern common building practice. Value tables were then created breaking down the value of the proposed plan based on the number of each type of mandate that it was in line with.

Minimize Waste Products Objective
Waste products are defined as the byproducts of source energy generation that cannot be recycled to generate more energy, and must be evacuated from the site. This is a concern for Army installations not only for its environmental implications, but also because of the cost associated with removing waste from post. The service might even be contracted to civilians, which would add to concerns surrounding the security of the installation. Data will be collected for this measure simply as a bulk volume of waste shipped away from the site (ft^3). As shown in Figure 4.3 we used bulk waste shipped as our value measure for waste products.

The Fort Carson report that we are basing our initial findings upon does not include any data for waste produced at the source energy sites. However, since we are striving to minimize waste products, any reduction in the amount from the baseline level is desirable, and will have value attached to it. Therefore, as the amount of waste collected decreases and we approach our target level of waste (ideally none), the more value is achieved.

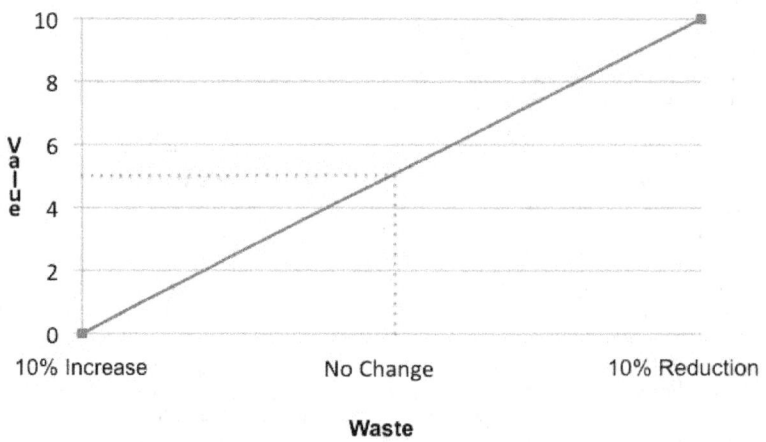

Figure 4.3 Value of waste from energy amount

Minimize Emissions Objective

Emissions are those waste products that are released directly into the air, and cannot be isolated and removed from the site. A good example of this might be smokestacks from a coal power plant, or the exhaust from the fleet of tactical vehicles described in the Fort Carson report. These are a particularly worrying byproduct of power generation, as they cannot or are economically unfeasible to collect. Emissions attributable to current source energy systems can be measured by volume (ft^3).

The Fort Carson report that we are basing our initial findings upon does not include any data for emissions produced at the source energy sites. However, since we are striving to minimize the amount of emissions, any reduction in the amount from the baseline level is desirable, and will have value attached to it. Therefore, as the amount of emissions decreases and we approach our target level (ideally none), the more value is achieved.

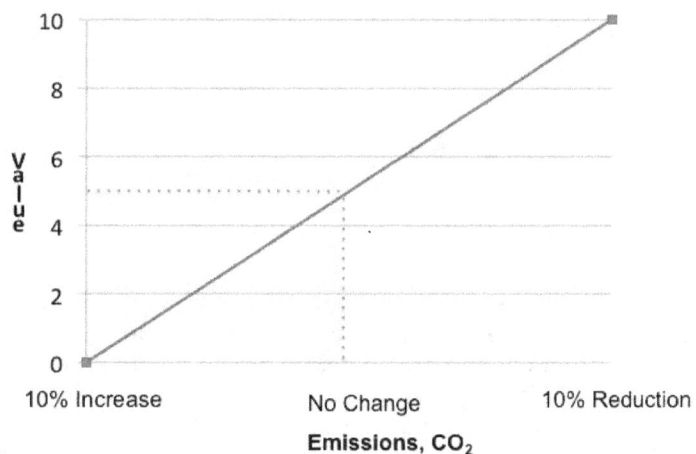

Figure 4.4 *Value of emissions from energy amount*

4.2.4 Increase Use of Renewable/Alternative Energy

Maximize Renewable Energy Consumption Objective

Energy consumption from different sources can be aggregated into BTU. This means that energy from photovoltaic cells and wind turbines, which are both variable, can be directly compared to the amount of energy generated from a nuclear power plant, which supplies power at more of a fixed capacity. Also, aggregating our data into BTU allows us to ignore differences in time as a dimension of other measures (i.e., instead of looking at kW/hr versus MW/hr, we can look at annual BTU). However, we must still differentiate between the sources of these BTU. In most cases it is currently neither physically, technically, or economically unfeasible to be able to generate 100 percent of an installations power requirements through renewable energy. As such, only a percentage of the total site energy will have come from these sources.

To illustrate a shift from fossil fuels to renewable energies, percent of site energy supplied by renewable sources can be measured. An increase in this percent is a desirable output, and as such, we will attach value to it. For every percent of power produced by renewable energy, one tenth of a point of value will be attached. In the example below, the target alternative is supplied by 77% renewable energy, thus it will have a value of 7.7 points.

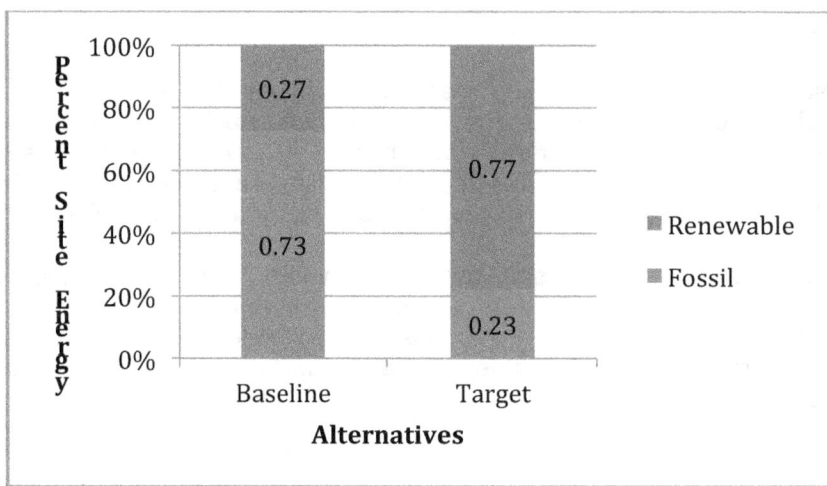

Figure 4.5 *Value measure for renewable energy consumption*

Note that the 77% could be representative of the objective energy or the minimum amount needed to conduct operations on the installation.

4.2.5 Assured Access to Sufficient Energy Supply
Maximize Resiliency Objective

Resiliency can simply be defined as the ability to bounce back from a disaster. This is presented pictorially in Figure 4.6 using a control theory perspective. From an energy security perspective, many scholars believe that we cannot protect our strategic resources from a determined threat. Instead our best defense is to build redundant and resilient systems. However, a review of the literature did not product any methodology to quantify resiliency. Thus, we propose the definition presented in Figure 4.7. This method utilizes the recovery time of different technologies in the aftermath of an attack. To this end, we are proposing that we measure the "Action" performed by these technologies. "Action" is defined as the amount of energy multiplied by the length of time that we will call Resiliency Factor. Note in the figure that two levels of energy needs are presented. The first is the "Typical Daily Requirements." The second is the "Minimum Amount to Accomplish Mission." Recent experience has shown that in the aftermath of a natural disaster that through energy conservation and other means that the populace requires significantly less to function. For example, Japan cut power consumption by 20 to 25% after the 2011 earthquake. With every military post being different, we suspect the ratio of the minimum divided by the typical to be on the order of 60%.

Figure 4.6 also contains the value function for resiliency. Note that the units of resiliency are MMBtu*Days. Fort Carson currently requires 2,616,402 MMBTu to operate[15] (NREL, 2010). If we assume 60% as the goal we need to return to 1,569,841 or roughly 1,570,000 MMBTu per year. If you assume that the minimum acceptable time is 30 days then the minimum acceptable amount (value of 0) is 30*1,570,000 or 47,100,000 MMBtu*Days/Year.

---

[15] This value does not include vehicle requirements.

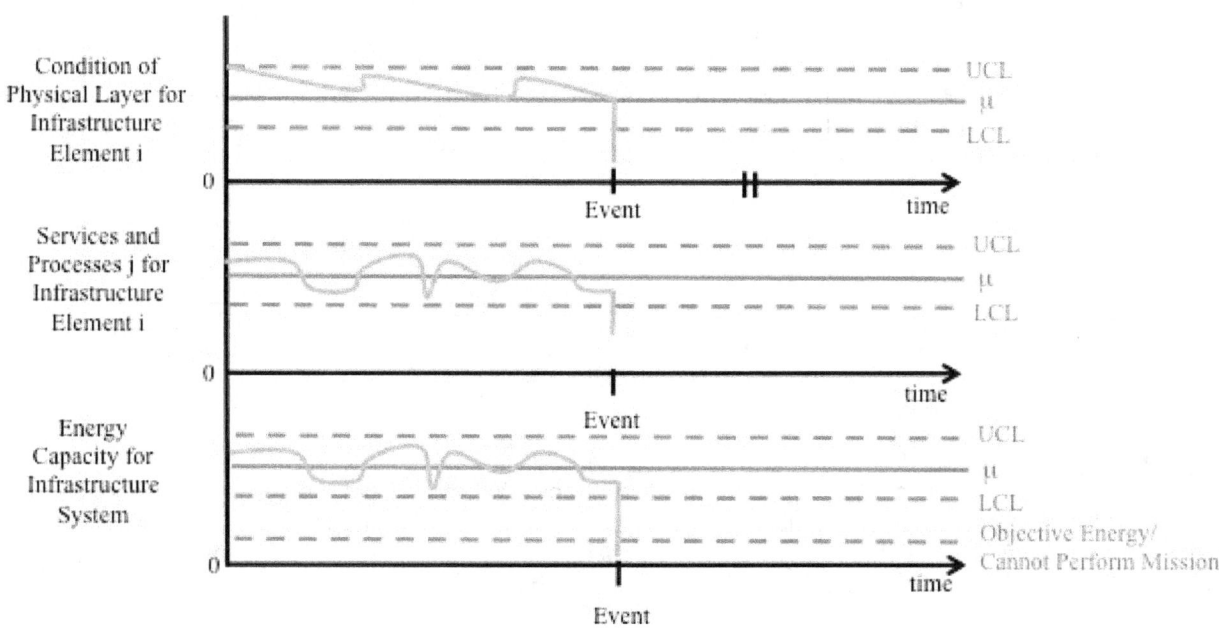

Figure 4.6 Control theory perspective of infrastructure and services used to describe how a significant event affects infrastructure and services on a military installation (modified from Goerger and Driscoll, 2006)

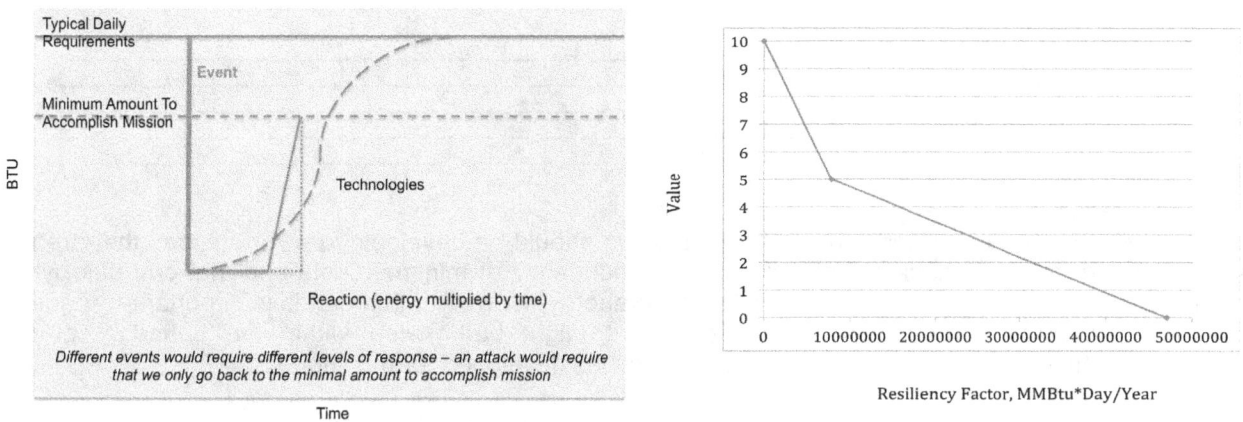

Figure 4.7 Graphical representation and value measure for resiliency

Note that when we discuss energy security (and not NetZero) that we are interested in objective energy or the amount of energy needed to sustain systems, information, and processes required to train, move, and sustain forces for military operations. Objective energy is the amount needed in the conduct of operations.

<u>Maximize Cyber Security Objective</u>
Our energy infrastructure is as at much or more risk from a cyber attack as a physical attack. All energy solutions depending upon the level of SCADA (supervisory control and data acquisition), exposure, complexity, etc., are all vulnerable to cyber attacked. For our value model we must consider the

vulnerability to cyber attack. We must protect the energy infrastructure from unauthorized access, disruption, modification, perusal, recording or destruction.

We chose to base our value model loosely on the confidentiality, integrity, and availability (CIA) model for information security. As shown in Table 4.4, we lumped confidentiality and integrity into general susceptibility value. We combined this with an impact value to measure both risk and outcomes.

Table 4.4 Value measure for cyber security

		Impact	
	Significant		*Minor*
Significant	0	2	5
Susceptibility	1	4	7
Minor	4	8	10

Maximize Physical Security Objective

Our energy infrastructure is also at risk from physical attacks. Whether from a disgruntled employee or an actual terrorist physical security is a major issue for all elements of the national energy grid. For our value model we must consider the vulnerability to a physical attack. We must protect the energy infrastructure from unauthorized access, disruption, modification, and/or destruction.

We chose to base our value model loosely on the confidentiality, integrity, and availability (CIA) model for information security. As shown in Table 4.5, we used a general susceptibility value. We combined this with an impact value to measure both risk and outcomes. Like cyber security, physical security has a high value scores for those technologies/solutions that are not susceptible nor have a major impact on the energy delivered.

Table 4.5 Value measure for physical security

		Impact	
	Significant		*Minor*
Significant	0	2	5
Susceptibility	1	4	7
Minor	4	8	10

Maximize Redundancy Objective

Many believe that in lieu of additional capacity that we should be developing power sources that are not only hardened from an attack but also maximize redundancy at minimal costs and that can quickly be restored. In some ways this also reflects alternative energy. However, this metric is important to tradeoff building redundancy versus alternative energy that might be located within the boundary of the installation. Table 4.6 shows that all of the value functions developed for this model.

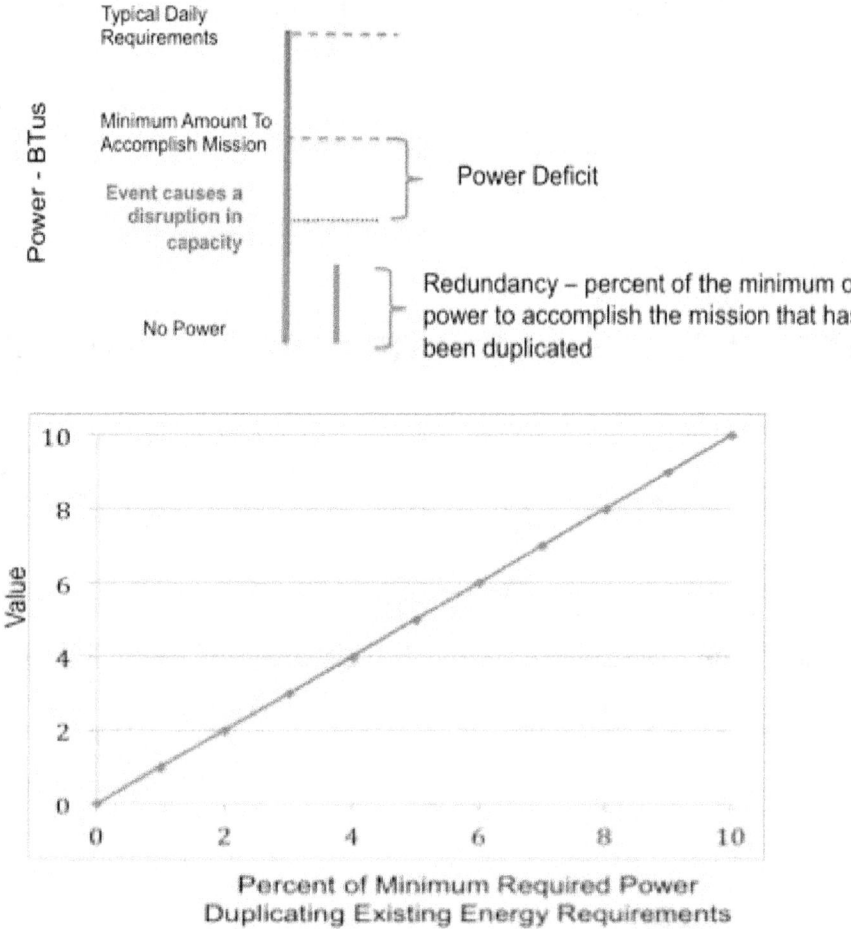

Figure 4.8 *Graphical representation of value measure of redundancy*

Table 4.6 *Value measures for the energy security and NetZero investment model*

Value Measure	*Comments*
Reduce Energy Consumption (ESG 1)	
	The amount of energy consumed will be measured by using the amount of site energy and source energy is used on the entire installation. It is necessary to reduce the **offsite energy consumption** because it makes the army reliant on an unprotected energy grid.
Increase Energy Efficiecny Across Platofrms (ESG 2)	
	Energy use will be measured using **energy use intensity** (EUI), which is calculated by energy consumed per year divided, by total floor space (kbtu/ft^2). This can be expressed as percent of the average energy use intensity for a similar building. Average energy use intensity is an important measure of building energy efficiency.
Increase Use of Renweable Alternative Energy (ESG 3)	
	To illustrate a shift from fossil fuels to renewable energies, percent of site energy supplied by renewable sources can be measured. An increase in this percent is a way to measure **renewable energy consumption.** For every percent of power produced by renewable energy, one tenth of a point of value will be attached. In the example, the target alternative is supplied by 77% renewable energy, thus a change in value of 1 corresponds to a 7.7% change in renewable energy consumption.
Assured Access to Sufficent Energy Supply (ESG 4)	
	Resiliency is the context of our problem is the time it takes for a military installation to return to a level of service needed to accomplish the mission. This includes infrastructure/physical layers and the services layer. Note that the units are time-energy. Note that objective energy is simply the minimum amount needed for the installation to conduct its mission.

	Impact		
	Significant		Minor
Significant	0	2	5
Susceptibility	1	4	7
Minor	4	8	10

All energy solutions depending upon the level of SCADA (supervisory control and data acquisition), exposure, complexity, etc., are all vulnerable to cyber attacked. For our value measure for **cyber security** we must consider the vulnerability. We chose to base our value model loosely on the confidentiality, integrity, and availability (CIA) model for information security. As shown, we lumped confidentiality and integrity into general susceptibility value. We combined this with an impact value to measure both risk and outcomes.

Table 4.6 *Value measures for the energy security and NetZero investment model (cont.)*

Value Measure	Comments

		Impact		
		Significant		Minor
Significant		0	2	5
Susceptibility		1	4	7
Minor		4	8	10

Our energy infrastructure is also at risk from physical attacks. Whether from a disgruntled employee or an actual terrorist **physical security** is a major issue for all elements of the national energy grid. We must protect the energy infrastructure from unauthorized access, disruption, modification, and/or destruction. As before, we chose to base our value model loosely on the confidentiality, integrity, and availability (CIA) model for information security. Like cyber security, physical security has a high value scores for those technologies/solutions that are not susceptible nor have a major impact on the energy delivered.

Many believe that in lieu of additional capacity that we should be developing power sources that are not only hardened from an attack but also **maximize redundancy** at minimal costs and that can quickly be restored. The corresponding figure shows that value function is determined.

Reduce Adverse Impacts on the Environment (ESG 5)

NetZero Significance Key	
1	Significant NetZero initiative
2	Minor NetZero initiative
3	Modern common building practices or Not applicable

Currently a near endless list of laws, mandates, executive orders, and regulations exists (around 70) in regards to the issues we are addressing in our study all designed to be in **compliance with mandates**. These mandates set a variety of goals and standards for military installations to meet. A portfolio or project can subjectively be categorized based upon one of the three categories presented.

Waste

Waste products are defined as the byproducts of source energy generation that cannot be recycled to generate more energy, and must be evacuated from the site. **Minimizing waste products** is a concern for Army installations not only for its environmental implications, but also because of the cost associated with removing waste from post. Note that maintaining the status quo does not contribute any value. Realist target levels must be set (some are set by directives/laws).

Emissions, CO$_2$

Emissions are those waste products that are released directly into the air, and cannot be isolated and removed from the site. A good example of this (although a disappointing visual) might be smokestacks from a coal power plant, or the exhaust from the fleet of tactical vehicles. These are a particularly worrying byproduct of power generation, as they cannot or are economically unfeasible to collect. **Minimizing emissions** attributable to current source energy systems can be measured by volume (ft^3).

4.2.6 Corresponding Swing Weights

As discussed in the methodology section, weights are needed for each associated value measure. The best means to prioritize and assign weights is to use a formal interview process and categorize the weights using a swing weight matrix. The value functions previously presented and there associated swing weights are shown in Table 4.7 below for energy security. We have also developed a swing weight matrix for NetZero initiatives, which is shown in Table 4.8. Note that Table 4.8 does not have a Maximize Resiliency value function. We felt that this not an appropriate metric for NetZero.

Table 4.7 Swing weight matrix for Fort Carson demonstration study for energy security

		Importance of the Value Measure to the Decision Makers and Stakeholders		
		High	*Medium*	*Low*
Variation in Measure Ranges	*High*	• Maximize Resiliency • Maximize Renewable Energy Consumption • Maximize Redundancy	• Minimize Energy Use Intensity • Maximize Compliance With Mandates • Minimize Emissions	
	Medium	• Minimize Offsite Energy Consumption		• Minimize Waste Products
	Low	• Maximize Cyber Security • Maximize Physical Security		

Table 4.8 Swing weight matrix for Fort Carson demonstration study for Netzero initiatives

		Importance of the Value Measure to the Decision Makers and Stakeholders		
		High	*Medium*	*Low*
Variation in Measure Ranges	*High*	• Maximize Renewable Energy Consumption	• Minimize Energy Use Intensity • Maximize Compliance With Mandates • Minimize Emissions	
	Medium		• Minimize Waste Products • Maximize Cyber Security • Maximize Physical Security	
	Low			• Minimize Offsite Energy Consumption

4.3 Life Cycle Considerations

Equivalent Annual Value[16] or EAV is a uniform flow of benefits less costs at equally spaced at equal time periods over the life cycle of the project. EAC is often used as a decision making tool in capital budgeting when comparing investment projects of unequal life spans. For example if project A has an expected lifetime of 7 years, and project B has an expected lifetime of 11 years it would be improper to simply compare the net present values (NPVs) of the two projects, unless neither project could be repeated.[17] It is a measure of the net return on a project on an annualized or amortized basis. EAV can also be calculated simply by converting the project NPV to an annuity or uniform cash flow. Because of government incentives, we chose to use year zero total costs. Because the life cycle of most renewable energies is roughly the same using year zero net present value is acceptable.

Developing actual costs of portfolios is complicated and probably worthy of its own research effort. Because of third party financing and government incentives, market conditions, etc., we did not have the information to develop EAV costs for our various portfolios. We used the Nation Renewable Energy Laboratory (NREL), Cost of Renewable Energy Spreadsheet Tool or CREST.[18] The CREST models described herein have been designed for use by state policy makers, regulators, utilities, beginning developers or investors, and other stakeholders. The models allow users to:

- Estimate the NPV cost of energy (COE) and levelized cost of energy (LCOE) from a range of solar, wind and geothermal electricity generation projects;
- Inform the process of setting of cost-based incentive rates;
- Gain understanding of the economic drivers of renewable energy projects, which lead to the calculated COE and LCOE; and

---

[16] Also referred to equivalent annual costs or EAC.
[17] Example was taken from Wikipedia at http://en.wikipedia.org/wiki/Equivalent_annual_cost
[18] The NREL CREAST model can be accessed at https://financere.nrel.gov/finance/content/CREST-model

- Understand the relative economics of generation projects with differing characteristics, such as project size, resource quality, location (e.g. near or far from transmission) or ownership (e.g. public or private).

4.4 Using Portfolios In Lieu of Projects

One thing becomes abundantly clear when trying to change energy supplies from fossil fuels to renewable sources. A major setback is that no single source of renewable energy is cost effective or reliable enough to raise the required amount of energy. If there was a single renewable energy source that could do that, it is likely the idea of a NetZero military base would be much easier to achieve. So, since one source is not enough, it is possible that an entire portfolio of renewable sources might help achieve the necessary production.

In an article written at Forsberg (2008), he explored this exact same concept of portfolio energy production, but at a larger scale. They consider in the article the idea of combining the three major forms of energy production: fossil fuels, renewable sources, and nuclear energy. Even now, these forms of energy production are considered competing sources. However, according to MIT, "new constraints and new technologies suggest that in many cases these energy sources must be tightly coupled to meet society's requirements" (Forsberg 2008). The article specifically cites electrical energy, which we are also specifically looking at, as this form of energy is what is needed to power buildings, lights, etc. The major challenge with electrical energy is the variance it has over the course of a year. Right now fossil fuel energy production struggles to keep up with peak energy requirement periods. If a portfolio concept was used, in some cased renewable sources would easily be able to take the pressure off. For example, energy production peaks when the sun is hottest in some areas in order to run air conditioning. At this same time, solar power would be able to produce at its highest levels as well, effectively leveling the requirement on fossil fuels (Forsberg 2008).

Table 4.6 shows three portfolios built based off the Fort Carson data. A single source would need to be massive and expensive to reach the levels of energy required. However, with the portfolio approaches NetZero as a combination of several energy producers, which are all on a much more manageable level. More portfolios can be built using this same concept.

By varying which producers are used and the capacity size of those producers each portfolio has a different "mixture" that can then be analyzed for security, cost, effectiveness, reliability, etc. The data used was based off data collected by the National Renewable Energy Laboratory at Fort Carson. Some assumptions were made in the creation of these portfolios. First, the energy requirement is specifically how much electrical energy projected needs in 2015, excluding other types of energy such as thermal energy and transportation energy. This energy requirement number also takes into account efficiency projects (i.e., "save a watt" campaigns, etc.) that would also likely cost money, but are not considered in this project. Second, there was an assumed interest rate of 8% in order to calculate the annual cost over each producer's life cycle. The life cycles were also assumed. Lastly, a linear correlation was assumed for adjusting capacity size and cost. The adjusted amount for the capacity is noted, and the costs are all increased by that same amount.

4.5 Demonstration Study Results

Based off the portfolios previously presented, a demonstration study was conducted using both models of analysis; multi objective decision analysis and data envelopment analysis. The goal of the study is to demonstrate the potential effectiveness of the analysis in determining the success or weakness of any portfolio or energy production plan created.

4.5.1 Multi Objective Decision Analysis

Looking at the efficiency frontier, the mixed portfolio comes out on top. Currently the mixed portfolio has the largest energy security value of 5.9 at an initial cost of only $1.243B[19]. From Figure 4.9 we can see our wind portfolios also contained similar value at significantly less cost. We can show from our analysis that the photovoltaic option is probably not feasible based upon the cost data we developed.

Looking from a Netzero perspective, the results are similar. Because our energy security and Netzero MODA models are heavily weighted on renewable energy we would suspect the relative values for energy security to mirror NetZero.

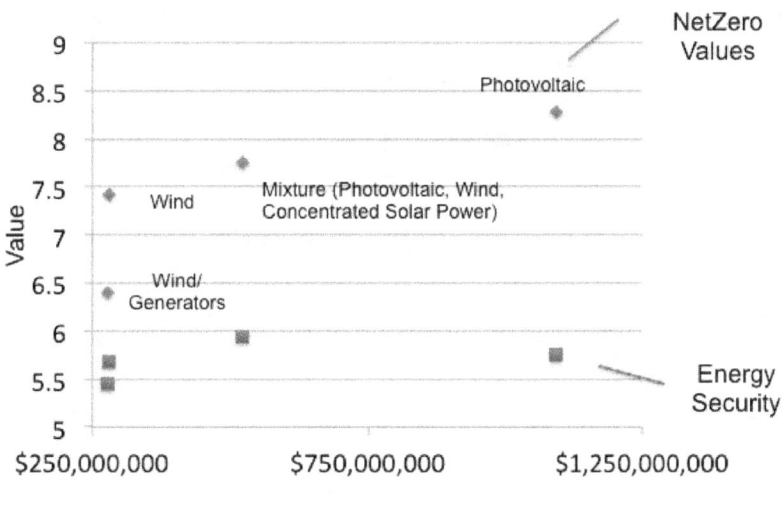

Figure 4.9 Graphical representation of cost versus value for energy security and NetZero

4.5.2 Data Envelopment Analysis

As previously discussed, DEA is another method used to analyze the most efficient energy portfolio, as well as identify the weaknesses of the other portfolios. For the demonstration study, all the portfolios were compared against each other, even though the portfolios are for either NetZero or for objective energy production. Since the DMUs for both types of portfolios are identical, DEA can still evaluate them and see which portfolios are most efficient. The inputs used included the total initial cost, the levelized cost of energy, and the outputs considered were the annual production of energy. Since this is a demonstration study, it is very likely that these numbers are not totally accurate, but the methodology can still show how DEA takes the information and instead of comparing the results to an objective value, but instead the actual achieved best results. Figure 4.10 shows the results of the efficiency calculation for all the portfolios.

This figure indicates that the wind and generator combined portfolios (Wing/Gen) for both NetZero and objective production is on the efficiency frontier, since its efficiency score is 1. This figure also helps visualize the severity of the difference between the other portfolios. For example, Photovoltaic's is far less efficient than Wind. This result seems to make since if you consider the different amounts of research and investments of different energy sources. It is very likely that as more research is done that the efficiency of some of the weaker portfolios will get better. Plus, as more portfolios are created that combine all the

---

[19] As previously discussed these costs values are for demonstration purposes. We did not have the total cost data for storage and other considerations.

renewable sources at different levels, it is also likely to improve efficiency of different renewable sources or even expand/increase the efficiency frontier entirely.

The efficiency frontier is one of the most basic guides for DEA. With the limited data an efficiency frontier is hard to build. Figure 4.10 shows the efficiency frontier for the portfolios. As a double input/single output analysis, the graph is based on the output divided by each input. This helps visualize the effect that each DMU has on the energy production, as well as make it possible to put in a two dimensional figure. DEA does not have to be limited to only a few measures in order to work. More DMUs can be created as long as they are consistent across all of the portfolios. However, once you have more than a few DMUs you will be unable to graphically depict the results of the efficiency frontier. The DMU indicates that the mixed portfolios are less than optimum a result that cannot be deduced from the MODA analysis.

Figure 4.10 Efficiency frontier

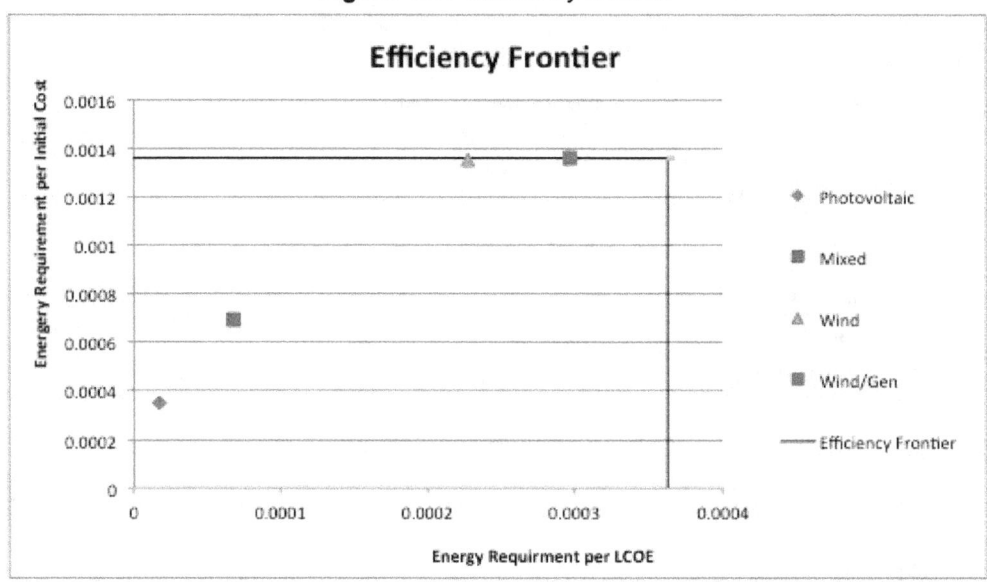

With more data, the frontier would not just be a box, but a curved line, so changing several DMUs would be the shortest distance to the frontier. Since the frontier is a box in this case, the necessary improvements are simple, either increase energy produced, or decrease the cost. This is easier said than done, but further DEA analysis can quantify exactly what amounts are required to reach maximum efficiency, as well as indicate when different renewable sources are worth the investment.

The recommendation based off the DEA analysis for this demonstration would be to recommend the Wind and Generator mixture to achieve both NetZero and energy security efficiency. However, more portfolios with different amounts of capacity would greatly help the results. This demonstration study shows the effectiveness of DEA, even with the limited amount of data. DEA could be even more effective in educating decision makers and stakeholders with more portfolios, more DMUs, and more analysis.

4.6 Summary

The demonstration study conducted took a few possible portfolios and analyzed them using MODA and DEA. The results would allow a recommendation to the stakeholder that could be easily supported. In this case, the value scoring supported the mixture portfolio, and the DEA provided support to the photovoltaic mixture. Leaving the final decision to the stakeholder, the demonstration allows value and analysis to help educate the decision and determine if the projects have a positive return. Any amount of portfolios and data could easily be placed in the same methodology and give effective results, just like the demonstration did.

This Page Intentionally Left Blank

Chapter 5
Summary

5.1 Conclusions

From this study, we can conclude that both MODA and DEA are effective ways of approaching this problem. Referencing the Systemigram presented earlier in this report, it is evident that MODA captures some – but not all – of the abstract externalities that must be considered. DEA provides an objective analysis of the purely technological components that our portfolios are comprised of, and ensure that we assemble portfolios that are both correct and efficient. However, neither can capture all of the aspects of this complex problem. The results from each of these different methods of analysis must be considered. This being the case, we would like to reinforce that our methodology does not provide a particular solution to the stakeholder. Instead, it presents information to the stakeholder in a manner that will allow them to make a more well-informed decision, whatever that may be. The stakeholder bears the ultimate responsibility of identifying which portfolio best fits the problem.

5.2 Future Work

Obviously, we would like to see our work applied in the future, but before this next step is taken, there are certain recommendations that we would like to make if this project were to be revisited. Firstly, more technological know-how will need to be incorporated in the development of portfolios. The portfolios that we generated in order to be able to demonstrate our methodology were developed solely by us; as Systems Engineers, we are not necessarily qualified to be able to determine what technologies are viable and which are not. Additionally, the knowledge of the technologies in place will greatly benefit our ability to predict the cost of these systems. Another idea that needs to be advocated is that the project team partner with a particular installation to implement a solution. We've already suggested that portfolios will need to be tailored to particular installations because of geography, proximity to population centers, etc. Developing an overarching solution to the problem at any level higher than the installation itself has proven to be exceedingly difficult. Our demonstration study of Fort Carson was generated off of numbers that might no longer be current, and without any cooperation from the staff of that post. A more cohesive partnership, with access to more accurate data (energy consumption, environmental considerations, etc.) would drastically benefit the quality of our report. Finally, more stakeholders need to be brought into the loop in this project. Not only does this help us to accurately reflect the weights of the different decision variables, but it also helps us to identify other areas of concern that we might not have thought to address. However, since this initiative lacks sufficient funding, it is not a priority for many commanders, and they have little incentive to buy in to the idea.

Note that Appendix A contains some additional research in the quantification of resiliency. This original research was conducted by Goerger and Driscoll (2006). We adapted this research as a means to measure resiliency. For our MODA model the simple definition proposed in Figure 4.7 will suffice. However, the work presented in Appendix A is a start to mathematically capturing the interdependences associated with the various categories that comprise infrastructure.

This Page Intentionally Left Blank

Chapter 6
Bibliography and References

Bibliography

Aimone, Michael A. "Cyber Warfare and the US Electric Grid Implications for National Security." Headquarters U.S. Air Force, 08 November 2010

Security, Energy, Environmental, and Encroachment (SEEE) Panel. "Installation 2025." U.S. Army, 23 July 2009

References

Anderson, Tim, "A Data Envelopment Analysis Home Page," http://www.emp.pdx.edu/dea/homedea.html#DEA_Title, accessed September 6, 2011

Army Science Board, "Installations 2025 Study Report". *Version 2.4*, pp 37-43, 2006

Army Senior Energy Council "Army Energy Security and Implementation Strategy", Pg 1, 13 January 2009

Brockhoff, K. "On the Quantification of the Marginal Productivity of Industrial Research by Estimating a Production Function for a Single Firm", German Economic Review Vol. 7, pp. 202-229, 1970

Chouhdry, Ali, "Improving Security at the Untied States Military Academy," United States Corps of Cadets, West Point, 28 April 2010

Charnes, A., W. Cooper, & E., Rhodes. "Measuring the efficiency of decision-making units," European Journal of Operational Research Vol. 2, pp. 429–444, 1978

Cooper, W.C., Seiford, L.M., and Tone, K., Data Envelopment Analysis, Norwell, Kluwer Academic Publisher, 2000

Department of Energy, Federal Energy Management Program, "Performing Energy Security Assessments - A How-To Guide for Federal Facility Managers," accessed at http://www1.eere.energy.gov/femp/pdfs/energy_security_guide.pdf, accessed December 15, 2011, January 2006

Forsberg, Charles, "Progress in Nuclear Energy," Elsevier Journal, Department of Nuclear Science and Engineering, Massachusetts Institute of Technology, http://mit.edu/ans/www/documents/seminar/F08/forsbergpaper.pdf, accessed 14 November 2011, 2008

Forsberg, Charles, Progress in Nuclear Energy, "Elsevier Journal". Department of Nuclear Science and Engineering, Massachusetts Institute of Technology, 2008, http://mit.edu/ans/www/documents/seminar/F08/forsbergpaper.pdf

Goerger, Niki C., and Driscoll, Pat, "Measuring Resiliency of Metropolitan Areas: A Systems Interdependency Framework," 74th MORS Symposium, 15 June 2006

Hope, Timothy, "A Value-focused Approach to Justify the Cost of Energy Security," Military Operations Research Society Workshop, 2010

Hughes, Larry "Quantifying energy security: An Analytic Hierarchy Process Approach," presented at the Fifth Dubrovnik Conference on Sustainable Development of Energy, Water, and Environment Systems in Dubrovnik, Croatia, September 2009, accessed at http://dclh.electricalandcomputerengineering.dal.ca/enen/2009/ERG200906.pdf, December 15, 2011.

National Renewable Energy Laboratory. "Targeting Net Zero Energy at Fort Carson: Assessment and Recommendations," U.S. Department of Energy, September 2010

Nautilus Institute for Security and Sustainable Development, "Synthesis Report for the Pacific Asia Regional Energy Security (PARES) Project, Phase 1 Framework for Energy: A Framework for Energy Security Analysis and Security Analysis and Application to a Case Study of Japan," June 9, 1998, Working Draft accessed at http://oldsite.nautilus.org/archives/pares/PARES_Synthesis_Report.PDF, December 15, 2011

Ramirez-Marquez, Jose Emmanuel, and Farr, John V., "Decision-making Approach for Catastrophic Scenario Selection in Disaster Recovery Planning," International Journal of Decision Support and System Technology, Volume 1, Number 2, pp 36-51, April - June, 2009

Appendix A
Measuring Resiliency of Metropolitan Areas:
A Systems Interdependency Framework[1]

Energy infrastructures, including national and regional electricity grids, form the backbone of the economy and enable the operation of other infrastructures such as telecommunications, transportation, emergency systems, wastewater treatment, and other basic services. Like our economy, the military cannot perform its mission without a resilient, secure, and robust energy infrastructure. This structure includes all suppliers, producers, and users along with any associated physical and information infrastructure, interfaces, processes, and support organizations. One of the keys elements of energy security is resiliency. This appendix presents a means to quantify resiliency in a manner that affords direct modeling and assessment for resource allocation and policy development purposes. Moreover, settling on a quantifiable definition of resiliency in comparison to robustness is a first step towards identifying effective risk management processes.

The 2009 National Infrastructure Protection Plan (NIPP) establishes DHS's risk management process. According to the NIPP, identifying the assets that comprise the nation's 17 critical infrastructure sectors and key resources within the National Asset Database represents the first step in its process. The 17 critical infrastructure categories used by DHS include:

- Energy,
- Transportation,
- Water,
- Public Health,
- Emergency Services,
- Telecommunications,
- Dams,
- Commercial Assets,
- Agriculture and Food,
- Banking and Finance,
- Defense Industrial Base,
- Nuclear Power,
- Postal and Shipping,
- Government Facilities,
- Information Technology,
- National Monuments and Icons, and
- Chemical and Hazardous Materials.

Absent from explicit consideration is a notion of information as an infrastructure component in a manner consistent with supply chain considerations.[2] Hence, we would add Information assets, including intellectual assets and service processes that are key elements of innovations, adaptations, maintenance, and deployment activities as part of the infrastructure landscape. This is an important inclusion for military infrastructure in which responses to opposing force actions depend on pre-positioned abilities to adapt,

---

[1] This work was originally developed and presented by Dr. Patrick Driscoll and Dr. N ki Goerger in a paper titled "Measuring Resiliency of Metropolitan Areas: A Systems Interdependency Framework," at the 74th MORS Symposium in June 2006. This appendix updates this methodology and extends it as another option for measuring resiliency of energy security.

[2] Lacy, S., F. Midgley, J. Riley, N. Williamson. 2011. "ITIL – Managing Digital Information Assets," Office of Government Commerce, The National Archives, United Kingdom.

and these lie principally resident in Information assets. Note that the systems and services considered relevant from a resiliency perspective are readily identifiable via a coherent stakeholder analysis and systems thinking[3] approach to identification and classification. The core set required for quantifying energy resiliency can likewise be developed as they are clearly a proper subset of the modified DHS categories. For clarity in what follows, we consider the modified DHS categories (18) in the context of a metropolitan infrastructure. Rural, remote, regional, and national settings and their corresponding infrastructure elements are similarly modeled.

Three key definitions are needed:
* **Definition**. *Resiliency* is the time required to return a critical set of infrastructure elements *I* to conditions *at least as good as prior* system condition states within normal control bounds after a system shock at time *t* has degraded them past their threshold levels.
* **Definition**. A set of infrastructure elements are primarily comprised of two subsets - a physical layer, x_i, and a service process layer, s_{ij}.
* **Definition**. *Robustness* is the number of critical infrastructure elements able to be maintained within normal control bounds above their threshold limits after a system shock has taken place.

We note that one might consider the properties of resilience and robustness separately for the two subsets. However, for infrastructure purposes, services are directly affected by the condition state of the physical layer. In the basic model that follows, we define a pairing between associated physical-service elements so as to directly model and isolate this quality of service (QoS) dependency in order to observe the dynamic effects of this linkage under shock and recovery conditions. The dynamic evolution of the model then propagates and makes accessible higher order effects.

Consequently, we define:

$i = 1, 2, ..., 18$	physical infrastructure elements
$j = 1, 2, ..., J$	critical infrastructure services (each physical element may facilitate more than 1 service)
$k = 1, 2, ..., K$	set of *i, j* physical-service pairs
$x_i(t)$	condition state of physical infrastructure element *i* at time *t*
$s_k(t)$	service process condition state for pair *k* at time *t*
$Q = [Q_{ik}(t)]$	state interactivity level matrix at time *t* (between physical elements and physical-service pairs)
$m_i(t^*)$	maintenance effects imposed on physical layer element *i* at time *t*
$m_k(t^*)$	maintenance effects imposed on physical-service pair *k* at time *t*

A basic, continuous time model of resiliency then takes the form:

$$\begin{bmatrix} x_i(t+1) \\ s_k(t+1) \end{bmatrix} = \begin{bmatrix} Q_{xx} & Q_{xs} \\ Q_{sx} & Q_{ss} \end{bmatrix} \begin{bmatrix} x_i(t) \\ s_k(t) \end{bmatrix} + \begin{bmatrix} m_i(t^*) \\ m_k(t^*) \end{bmatrix} \quad \text{(A.1)}$$

which expresses the condition state vector $[\mathbf{x} \quad \mathbf{s}]^T$ at any time in the planning horizon. In doing so, $[\mathbf{x}(0) \quad \mathbf{s}(0)]^T$ represents an initial state condition to which the system desires to return in the event of a shock. This assumption can be changed in the model to accommodate alternative system states. In (A.1), we assume the dynamics matrix Q is time invariant, asymmetric and stable in basic model. This matrix Q is used to introduce dynamic change to system state levels caused by normal deterioration over time as well as shocks. The block structure of Q represents the pair-wise influence effects between state

---

[3] Jackson, M. 2003. *Systems Thinking: Creative Holism for Managers*, John Wiley & Sons, New York.

components as a result of linked interdependencies. The vector $[m_i \quad m_k]^T$ represents planned periodic maintenance effects to improve state conditions over a (possibly unique) periodic time sequence.

This basic model can be used to reveal insights to inform prioritized effective investments during 'normal' operating conditions. Figure A.1 shows our basic model.

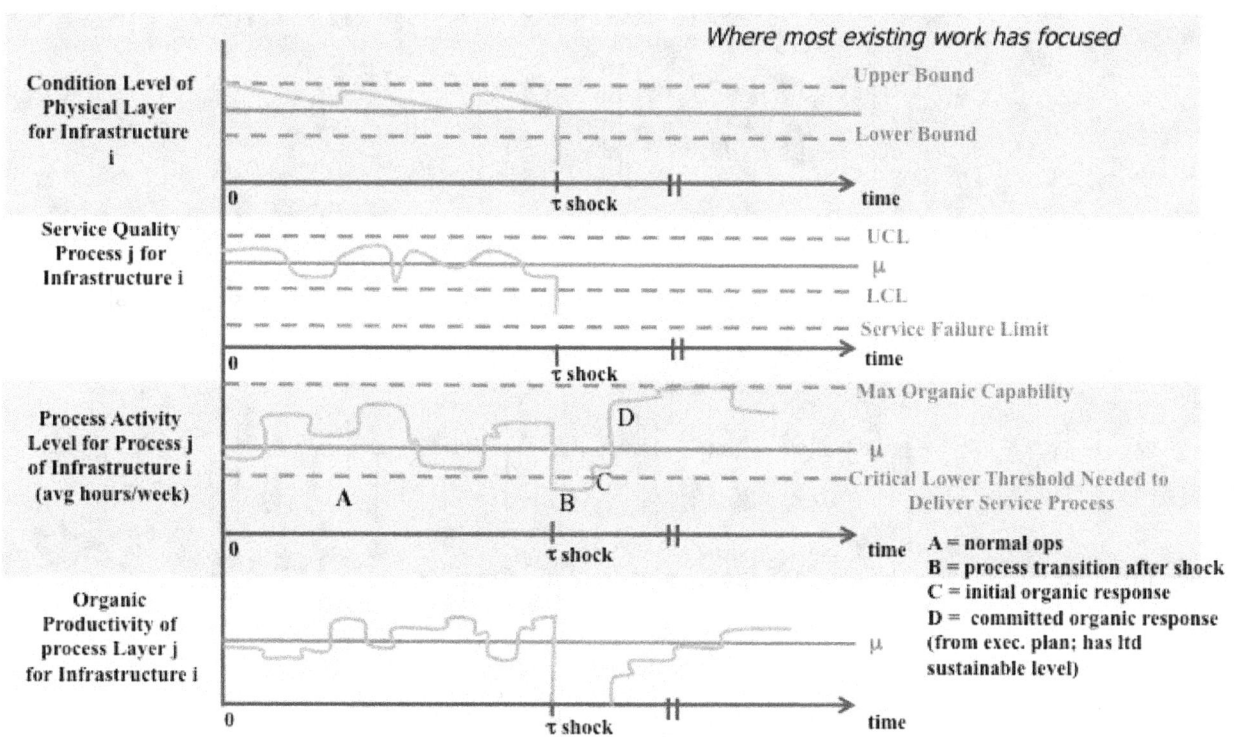

Figure A.1 Problem definition diagram

With a system shock occurring at time $t = \tau$, the basic model becomes:

$$\begin{bmatrix} x_i(t+1) \\ s_k(t+1) \end{bmatrix} = \begin{bmatrix} Q_{xx} & Q_{xs} \\ Q_{sx} & Q_{ss} \end{bmatrix} \begin{bmatrix} x_i(t) \\ s_k(t) \end{bmatrix} + \begin{bmatrix} m_i(t*) \\ m_k(t*) \end{bmatrix} + \begin{bmatrix} c_i(\tau) \\ c_k(\tau) \end{bmatrix} \qquad (A.2)$$

where $c_*(\tau)$ are characteristic effects imposed on the infrastructure based on specific scenarios being explored. Shock effects can also affect the dynamics contain in Q by degrading or amplifying interdependencies at $t + \varepsilon \cong t$, where $\varepsilon \ll$ incremental time step in model. Hence,

$$\begin{bmatrix} x_i(t+1) \\ s_k(t+1) \end{bmatrix} = \begin{bmatrix} \tilde{Q}_{xx} & \tilde{Q}_{xs} \\ \tilde{Q}_{sx} & \tilde{Q}_{ss} \end{bmatrix} \begin{bmatrix} x_i(t) \\ s_k(t) \end{bmatrix} + \begin{bmatrix} m_i(t*) \\ m_k(t*) \end{bmatrix} + \begin{bmatrix} c_i(\tau) \\ c_k(\tau) \end{bmatrix} \qquad (A.3)$$

where

$$
\begin{bmatrix} \tilde{Q}_{xx} & \tilde{Q}_{xs} \\ \tilde{Q}_{sx} & \tilde{Q}_{ss} \end{bmatrix} = \left(\begin{bmatrix} Q_{xx} & Q_{xs} \\ Q_{sx} & Q_{ss} \end{bmatrix} + \begin{bmatrix} C_{xx}(\tau) & C_{xs}(\tau) \\ C_{sx}(\tau) & C_{ss}(\tau) \end{bmatrix} \right) \tag{A.4}
$$

Thus, dynamic effects are defined and introduced into the infrastructure by way of triplets:

$$
\left(\begin{bmatrix} C_{xx}(T) & C_{xs}(T) \\ C_{sx}(T) & C_{ss}(T) \end{bmatrix}, \begin{bmatrix} c_i(T) \\ c_k(T) \end{bmatrix}, T = \{t_1, t_2, \ldots\} \right) \tag{A.5}
$$

in which T is unitary in this model.

Responses to shocks can likewise be integrated into the basic model. Assuming that a system shock occurs at time $t = \tau$, and that preplanned responses occur at $t = \delta$, intended to restore/stabilize the dynamics of Q upset by the shock can be expressed as:

$$
\begin{bmatrix} x_i(t+1) \\ s_k(t+1) \end{bmatrix} = \left(\begin{bmatrix} \tilde{Q}_{xx} & \tilde{Q}_{xs} \\ \tilde{Q}_{sx} & \tilde{Q}_{ss} \end{bmatrix} + \begin{bmatrix} R_{xx}(\delta) & R_{xs}(\delta) \\ R_{sx}(\delta) & R_{ss}(\delta) \end{bmatrix} \right) \begin{bmatrix} x_i(t) \\ s_k(t) \end{bmatrix} + \begin{bmatrix} m_i(t^*) \\ m_k(t^*) \end{bmatrix} + \begin{bmatrix} c_i(\tau) \\ c_k(\tau) \end{bmatrix} + \begin{bmatrix} r_i(\delta) \\ r_k(\delta) \end{bmatrix} \tag{A.6}
$$

in which the response plan is likewise defined by triplets of the form:

$$
\left(\begin{bmatrix} R_{xx}(\delta) & R_{xs}(\delta) \\ R_{sx}(\delta) & R_{ss}(\delta) \end{bmatrix}, \begin{bmatrix} r_i(\delta) \\ r_k(\delta) \end{bmatrix}, \delta = \{t_1, t_2, \ldots\} \right) \tag{A.7}
$$

Here, the matrix R represents planned effects to restore/stabilize the dynamics of Q upset by the shock. The vector $[r_i \quad r_k]^T$ represents targeted effects to increase state condition levels. These then allow for the possibility of such targeted effects occurring at planned or periodic times subsequent to the shock.

Overall then, an appropriate policy plan in the context of this basic model consists of shock-response sets intended to address specific scenarios affecting the physical layer, or the service layer, or the dynamic dependencies existing between these elements:

$$
\left(\begin{bmatrix} c_{xx}(T) & c_{xs}(T) \\ c_{sx}(T) & c_{ss}(T) \end{bmatrix}, \begin{bmatrix} c_i(T) \\ c_k(T) \end{bmatrix}, T = \{t_1, t_2, \ldots\} \right) \tag{A.8}
$$

$$
\left(\begin{bmatrix} R_{xx}(\delta) & R_{xs}(\delta) \\ R_{sx}(\delta) & R_{ss}(\delta) \end{bmatrix}, \begin{bmatrix} r_i(\delta) \\ r_k(\delta) \end{bmatrix}, \delta = \{t_1, t_2, \ldots\} \right) \tag{A.9}
$$